Alonzo T. (Alonzo Thrasher) Keyt

Sphygmography and Cardiography

Physiological and Clinical

Alonzo T. (Alonzo Thrasher) Keyt

Sphygmography and Cardiography
Physiological and Clinical

ISBN/EAN: 9783337274559

Printed in Europe, USA, Canada, Australia, Japan

Cover: Foto ©berggeist007 / pixelio.de

More available books at **www.hansebooks.com**

SPHYGMOGRAPHY

AND

CARDIOGRAPHY

PHYSIOLOGICAL AND CLINICAL

BY

ALONZO T. KEYT, M.D.

EDITED BY

ASA B. ISHAM, M.D.,

AND

M. H. KEYT, M.D.

NEW YORK AND LONDON
G. P. PUTNAM'S SONS
The Knickerbocker Press
1887

EDITORS' PREFACE.

ALONZO THRASHER KEYT, M.D., died suddenly, without premonition, from paralysis of the heart, on the 9th of November, 1885, in the fifty-ninth year of his age. The post-mortem examination revealed calcareous granules in the segments, with incompetency of, the mitral valve, atheromatous degeneration in the heart and aorta, fatty and friable condition of the cardiac muscles, and dilatation of the left cavities. These conditions pertain to advanced age, and Dr. Keyt was aged prematurely from overwork. Besides attending to the demands of a large practice, he pursued unremittingly, from the commencement, his studies of the graphic method as applied to the circulatory apparatus, and in this labor of his love and life he was insensible alike to the suggestions of hunger or fatigue. He was a patient, careful, conscientious worker, seeking only in his investigations to arrive at the truth. He put nothing forth hastily or upon impulse but only after mature deliberation, subjecting his conclusions to the most rigid scrutiny, and sparing no time or trouble to test their accuracy by all the means within his knowledge.

His life-work is ended, but its fruits remain. These are scattered throughout various medical journals in irregular contributions. It seemed fitting that this valuable material should be placed in more convenient and compact form, for the common benefit of the profession, and as a memorial of one who might well be selected as a model of the American physician and of all that constitutes a true man. Accordingly it has been arranged in book form and in some order of sequence. Nothing more is assumed on the part of the editorial supervision than that it presents Dr. Keyt's work, in an available shape, to the profession.

A. B. I.

WALNUT HILLS, CINCINNATI, O.,
Nov. 9, 1886.

CONTENTS.

INTRODUCTION.

History of the Sphygmograph—The Sphygmometer—Keyt's Compound Sphygmograph: Description and Management—Water and Air as Media of Transmission for Sphygmographs 1

PART I.

PHYSIOLOGICAL SECTION.

CHAPTER I.

Form, Relationship, and Chronometry of the Cardiac and Arterial Movements—The Velocity of the Pulse Wave and Duration of the Ventricular-Carotid and Ventricular-Presphygmic Interval in Young Children 29

CHAPTER II.

Cardiography—Ventricular Systole and Diastole: Their Relation to Each Other and the Heart's Revolution—Mechanism of the Cardiac and Arterial Traces—The Negative Arterial Trace—The Presphygmic Interval 53

CHAPTER III.

PULSE-WAVE VELOCITY.

An Experimental Inquiry into the Causes of the Variations of Pulse-Wave Velocity 83

CHAPTER IV.

PULSE-WAVE VELOCITY (*Continued*).

Facts and New Experiments in Illustration of the Variations of Pulse-Wave Velocity in Man, and Bearing upon the Elucidation of the Causes Which Produce Them 104

CHAPTER V.
THE PRESPHYGMIC INTERVAL.

The Causes of the Variations of the Cardio-Aortic or Presphygmic Interval 120

CHAPTER VI.

Influence of Muscular Exercise on the Arterial and Cardiac Pulsations—The Pulsations of the Fontanel: Their Form and Mechanism and Relation to the Pulsations of the Heart and Arteries and the Movements of Respiration 130

PART II.
CLINICAL SECTION.
PRELIMINARY.

Variations of the Presphygmic Interval in Disease, Illustrated by Experiments upon the Schema 151

CHAPTER VII.

Diminution of Retardation of the Pulse in Aortic Insufficiency—The Influence of Aortic Aneurism and Aortic Insufficiency, Singly and Combined, on the Retardation of the Pulse 159

CHAPTER VIII.

Retardation of the Pulse in Mitral Insufficiency—Enormous Retardation of the Pulse from Mitral Insufficiency, Aortic Aneurism, and Heavy Aortic Valves 177

CHAPTER IX.

Cardio-Sphygmographic History of Aortic Obstructive Lesions—The Sphygmographic Indications in Aneurism 198

CHAPTER X.

The Cardio-Sphygmography of Tricuspid Lesions—New Interpretation of Flint's Mitral Direct or Presystolic Murmur without Mitral Lesions 219

INTRODUCTION.

History of the Sphygmograph—The Sphygmometer—Keyt's Compound Sphygmograph—Description and Management—Water and Air as Media of Transmission for Sphygmographs.

THE circulation of the blood, in its maintenance of or departure from the healthy standard, affords most important indications of the state of the system and its organs. The pulsations of the heart and accessible arteries express to our senses these indications. *Feeling the pulse* is an ancient and ever-surviving art, and the possession of the *tactus eruditus* has always and justly been a distinguishing qualification of the good physician. For the idea of studying and utilizing the pulse in other ways than by the sense of touch alone, viz.: to represent it to the eye and obtain its graphic line by means of an automatic registering apparatus, history carries us back to the immortal Galileo, who, it is said, among the many wonderful things that he did, constructed also a " pulsilogia." There is no description of the mechanism of this instrument now extant, and no other information respecting it is at hand than that it exhibited and recorded the movements of the pulse. It does not seem to have brought the inventor under the ban of the clergy, and therefore justifies the conclusion that it was considered a thing of no possible consequence for the time being or to come, and, as a matter of course, unworthy of any notice or special comment.

The graphic method thus originated was covered over with the mould of centuries, and did not again become applicable to the study of the circulation until the genius of Vierordt produced the sphygmograph. Vierordt's instrument consisted essentially of a spring, which was applied to the artery, and a lever connected with the spring, which traced the arterial movements upon a revolving cylinder covered with paper. The mechanism was clumsy and complicated, the tracing furnished only regular ascending and descending lines, the sole variation in the different strokes consisting in the curve, which was greater or less in accordance with the length of the pulsation. On account of these drawbacks the instrument did not meet with marked approval of investigators, or materially enhance the existing

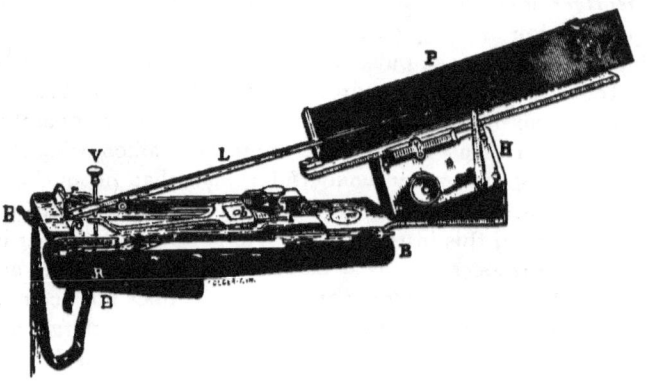

FIG. 1.

knowledge respecting the problems involved in the current of the circulating blood. It was, however, constructed upon correct principles, and needed only the skilful touch to convert it into an instrument of precision.

In 1860 M. Marey constructed the famous sphygmograph which bears his name. It is the same in principle as Vierordt's, but greatly simplified. It correctly records the movements of the arterial vessels under the varying pressure of the blood column within them. It has contributed many important facts concerning the physiology and pathology of the circulatory apparatus, and the great master, under whose hand it was fashioned into a working instrument, has, through its aid, established a scientific method of investigation, which forms a landmark in the line of human progress, and holds out to earnest workers the richest prospects of future rewards. Since the introduction of Marey's instrument it has been modified in various ways, but it remains essentially the same. The work it has performed under the guidance of Marey, Wolf, Lorain, Burdon-Sanderson, Franck, and others is, in the aggregate, immense.

In 1874 Dr. A. T. Keyt devised a sphygmometer, at that time unaware that M. le Docteur Hérrison, in 1835, had constructed an instrument similar in principle but different in mechanism, and for which he had been derided by the editor of the *Medico-Chirurgical Review*, as having produced a bauble meriting only the contempt of those who possessed the cultivated touch. The sphygmometer of Dr. Keyt, as shown in the cut, consisted of a base or receptacle, *a*, made of thin brass, semi-circular in form, with an oblong free edge below, and a shallow neck, *b*, above, into which is inserted, air-tight, the glass tube *c*. The free edge of the base measures inside one inch and three-eighths in length by three-eighths in width; and over it is drawn a rubber membrane, air-tight, and just tense enough to secure its smoothness and integrity of action.

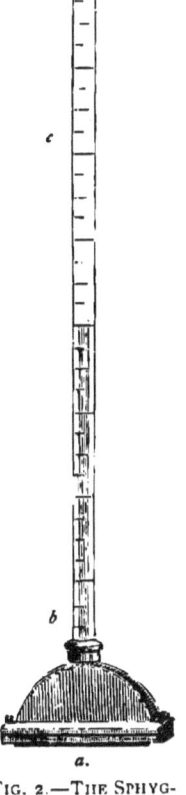

Fig. 2.—The Sphygmometer.

The glass tube is eight or ten inches long, of small

bore, and graduated in inches, halves, and quarters. The base filled with water to the top of the neck, the instrument is ready for its purpose.

With the base pressed directly over an artery, as the radial, the elastic floor expands, closes round and accurately fits the vessel as the segment of a sheath, the liquid in the meantime rising in the tube in proportion to the expansion upward of the basic membrane. The elastic coat so embracing the artery will move exactly with its movements. There is, however, a degree of pressure at which the tension of the membrane so counterpoises the tension of the artery as best to develop these movements. This is shown by the undulations of the liquid column, which are true and exact pulsations of the artery transferred to the liquid in the tube. By placing a rubber membrane over the end of the glass tube and superimposing thereon a pin and lever, as shown in the cut, the movements of the liquid column became further transmitted. It now only needed a registering apparatus, and Dr. Keyt's first sphygmograph was developed as exhibited on p. 5.

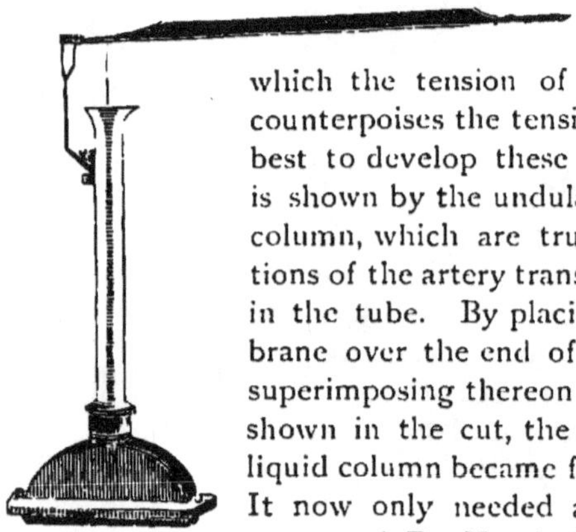

FIG. 3.

This instrument (Fig. 4) was used and improved, but it was soon found that a single tracing offered little of value for clinical or physiological purposes. Accordingly a double sphygmograph was constructed, and, in 1876, Keyt's compound sphygmograph, or cardiograph and sphygmograph combined with chronographic attachment, as shown in Fig. 5, was made available for the study of the circulation. With this instrument most of the work embodied in the pages of this book was performed. (See Fig. 5, page 6.)

A few years later the perfected mechanism exhibited in Fig. 6 was attained, and all of Keyt's compound sphygmographs are now constructed after this model.

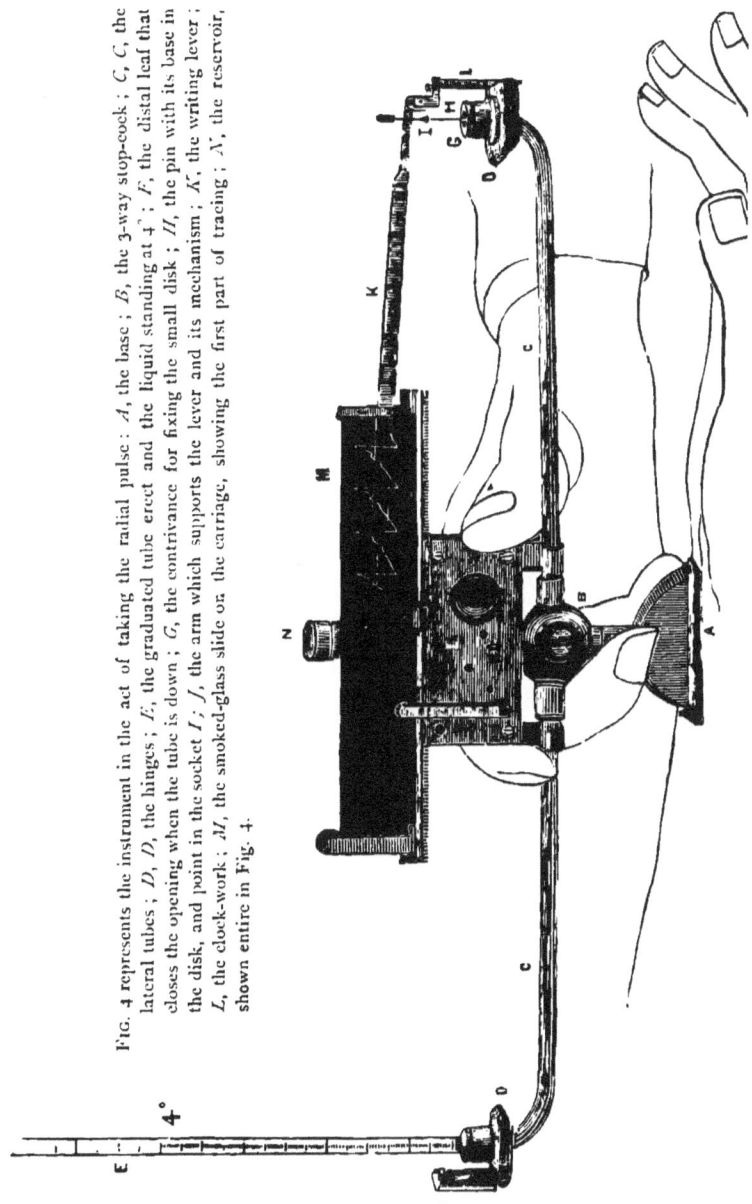

FIG. 4 represents the instrument in the act of taking the radial pulse: *A*, the base; *B*, the 3-way stop-cock; *C, C*, the lateral tubes; *D, D*, the hinges; *E*, the graduated tube erect and the liquid standing at 4°; *F*, the distal leaf that closes the opening when the tube is down; *G*, the contrivance for fixing the small disk; *H*, the pin with its base in the disk, and point in the socket *I*; *J*, the arm which supports the lever and its mechanism; *K*, the writing lever; *L*, the clock-work; *M*, the smoked-glass slide on the carriage, showing the first part of tracing; *N*, the reservoir, shown entire in Fig. 4.

6   THE SPHYGMOGRAPH.

C H is the chronograph, of which *W'* is the winder, *D'* the dial, *P'* the pendulum, and *L'* the tracer ; *Z* is the double rest for the levers. On the glass, in direct view, are three tracings —one of the heart, one of an artery, and one of time in seconds and fifths.

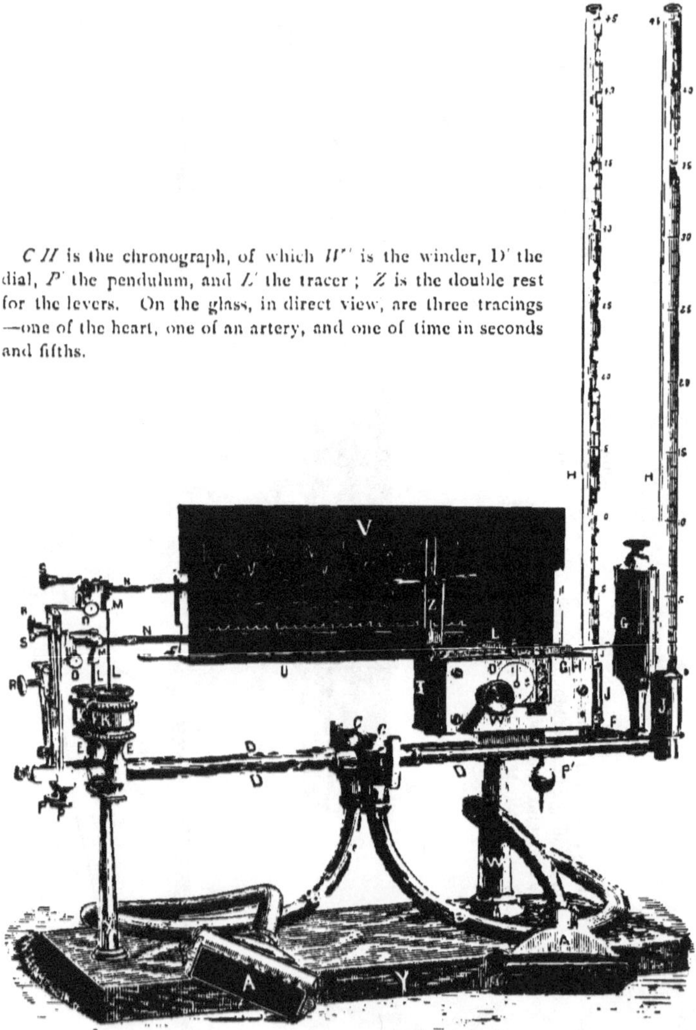

FIG. 5.—KEYT'S COMPOUND SPHYGMOGRAPH.

INTRODUCTION.

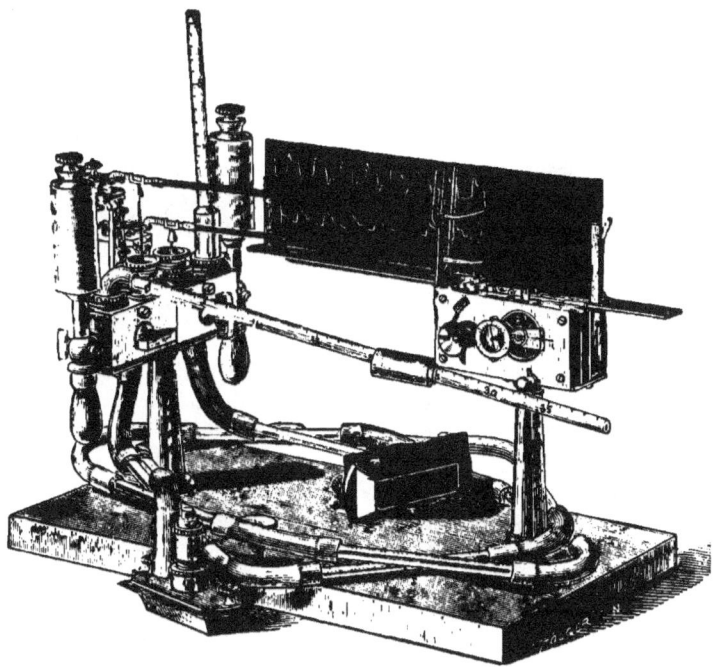

FIG. 6.

DESCRIPTION AND MANAGEMENT OF KEYT'S COMPOUND SPHYGMOGRAPH.

In order to successfully use the compound sphygmograph, some preliminary study of its mechanism and practice in its management are requisite. However, the instrument is simple and easily understood, and skill in its use is soon acquired. The following instructions will be of advantage to those who would comprehend and employ the apparatus.

The Basal Membrane.

The basal membrane is of pure-rubber cloth, and is secured air-tight over the bottom of the receptacle by means of the close-fitting exterior rim ; the latter screwed home by thumb-screws, one at either end. The proper degree of tension of the basal membrane is secured by means of an interior rim or " tensor," which is protruded against the membrane by direct turns of the upper of the two screw-thimbles which encircle the central stem, the latter being screwed into a central bar of the adjusting rim or tensor. The lower thimble secures the packing around the stem. By this arrangement the basal membrane can be readily placed and kept at standard tension.

Transmission Tubes.

The transmission tubes are of glass with stiff-rubber joinings, twenty-five inches long, one-eighth to one-sixth-inch bore, and always uniform with each other.

The 3-Way Stop-Cock.

The 3-way stop-cock is placed between the circular chambers. When the key is horizontal with the marked face down, communication is free between the transmission tube and both circular chambers ; when the key is vertical with marked face towards the index tube, communication is open only between the transmission tube and index chamber; when the key is vertical with marked face towards the lever, communication is open only between the transmission tube and lever chamber ; lastly, when the key is horizontal with marked side up, communication is open only between the two chambers.

Index Tube.

The manometric or index tube is graduated in conformity to a standard ; and is secured over its chamber by the device proper thereto. It is an indispensable part of the apparatus and has several important uses.

The Small Membrane.

The small membrane with vertical pin is secured over its chamber as follows: on the shoulder above the chamber is

INTRODUCTION.

first smoothly placed a rubber ring cut for the purpose, then the membrane with pin attached, then another rubber ring, then a metal ring, and then over all the fitting thimble is screwed firmly down. The small membrane is brought to its proper tension by means of a thimble shod with a circular tensor, and which screws down until the tensor presses sufficiently against the membrane, and then a check-ring running on the circumference of the thimble is screwed down to make all solid.

The Upper Stop-Cock—

Opens when horizontal, and closes when vertical, a passage between the upper part of the circular chambers.

The Reservoir—

With its ampulla and 3-way stop-cock conveniently provides for the supply of water to the apparatus.

The Lever—

Moves freely on its axis, which latter is pivoted to an adjustable plate. A small hollow cone attached to a clip that slides on the lever lodges the point of the pin, on which the lever plays; and the lever and cone are adjusted so that one half inch separates the apex of the cone from the centre of the axis. The lower thumb-screw fixes the lever in its longitudinal position; and the upper thumb-screw adjusts the level of the lever. Also a very delicate screw placed on the lever adjusts its point sideways so as to bring it in proper contact with the writing surface.

The Filling.

With a pipette fill the reservoir partly full and close the stop-cock towards the chambers; open the other stop-cocks; detach the base from the transmission tube; place base under water with its opening uppermost, and then with the finger work the membrane until air bubbles no longer escape; place the end of the transmission tube in water, then take the end of the index tube in the mouth and suck up the water until a solid column traverses the whole length without air bubbles; reverse the 3-way stop-cock, which will hold the water from escaping, then join the base and transmission tube under water; open the 3-way stop-cock towards the lever chamber and work the basal membrane with the finger, turning the instrument on its side, to favor escape of air from the lever chamber, if any has been retained there; turn the instrument back and close the upper stop-cock, and by means of the ampulla and stop-cock of the reservoir adjust the water so as to stand at 0° with the base lying on the table.

(*Use no liquid but pure water.*)

Adjustment of the Membranes.

The basal membrane is adjusted to standard tension as follows: With the base lying on the table and water in the index-tube marking 0°, the upper and reservoir stop-cock shut off, open the 3-way stop-cock only towards the index chamber, then elevate the base to the full height permissible, and note the ascent of liquid in the index tube; repeat the trials and vary the tension until the liquid under the conditions rises to 18 degrees, which has been adopted as the standard. If to obtain the standard the membrane requires too great protrusion of the tensor, withdraw the latter, unscrew the outer rim, and replace the membrane with less slackness; a small protrusion of the tensor should effect the object.

The small membrane is adjusted to standard tension by the following formula: With the instrument prepared as for adjustment of the basal membrane, open the 3-way stop-cock only towards the lever chamber, then elevate the base as before described, and note the elevation of the point of the lever; adjust the tension of the small membrane until the lever rises under the conditions seven tenths ($\frac{7}{10}$) of an inch, which is the standard adopted as yielding the most satisfactory results.

When desired to limit the excursion of the lever, as may sometimes be the case in tracing hypertrophied hearts, this is readily done by increasing the tension of the small membrane; but as soon as the particular experiment is over, the membrane should be restored to its standard tension.

As the membranes relax from use they will require testing and adjusting from time to time; and as soon as they show leakage or failure of sensibility, they must be replaced by new ones.

The Registering Part.

The clock-work movement is wound up by the proper winder. The carriage is entered and removed sideways by pressing out the spring-groove. The glass slide properly smoked is placed in the carriage, and more conveniently before the latter is placed in the grooves. The clock-work is started and stopped by the following device:

The Air Trigger.

In order that the hands may be free, a device provides for starting and stopping the movement by the action of the mouth on the air contained in a rubber tube. Compressing the air pushes out a membrane contained in a small metal capsule, carrying with it a pin attached to its centre, the latter striking against the works and stopping the movement; exhausting the air withdraws the membrane and pin, freeing the works and starting the movement.

The Chronograph.

The chronograph is wound up by the winder on its face, and started and stopped by turning the small knob, towards the opposite end. Its writing lever is raised every time an arm of the escapement lever passes its centre, which is five times every second, and therefore the chronograph writes fifths of seconds. A small eccentric serves as a rest for the writing lever, and regulates by its position the height of the points. A small elevation is preferable. The dial hand makes a revolution per minute, and the correctness of the chronograph can be tested by comparing this hand with the seconds hand of a watch. The chronograph must be placed so that its writing point just lightly touches the slide. By means of the chronograph the time of the movements and the time intervals between movements recorded can be measured with precision.

How to Use and Manage the Instrument.

To use the instrument, for instance to trace the pulse, turn the 3-way cock so as to open only towards the index tube, and place the base lengthwise on the artery in such manner and with such pressure as will develop the highest undulations of the liquid column. Note the height of the undulations by the degree marks, and the number of degrees the column rose above the starting lever. The first measures the amplitude of the pulse, and the last the pressure required to develop it. Next, whilst the liquid column is displaying its best movements, holding the base steady in position, turn the cock so as to open only towards the lever, and then use a pressure that will develop the highest sweep of the lever, and whilst the latter is executing its best movements start the carriage and take the tracings.

Other movements, as the pulsation of the heart, are taken much in the same manner; and simultaneous tracings of two movements are obtained by placing the bases on the respective parts and receiving the inscriptions on the same slide. The chronograph is always kept running during the experimenting, writing the exact time of the movements inscribed.

When using the instrument from time to time, keep it well charged in the intervals of non-use, and with the 3-way stop-cock open in the three directions. When the instrument is not to be used for a considerable time, it is better to remove the membrane and put it away with the interior dry.

When not in use, the index tubes may be turned down or kept erect, as convenience may dictate. It is well to have at hand duplicate glass tubes to provide against accidental breakage. Always keep a supply of glass slides cut from thin glass with even surface; reject slides which present a curve. Let the slides be evenly and not too heavily smoked.

After simultaneous tracings have been obtained repass the slide, halting it at selected points, and strike the levers, by pressing the basal membrane with the finger, across the line of traces, to obtain signal lines which will indicate the time relations to each other of the movements inscribed. The difference in distance of the beginning of the up-strokes from these lines, measured on the corresponding chronogram, gives the exact time which separates the two movements from each other. If the movements are synchronous the signal lines will cross the respective traces at exactly the same stage.

When desired, photographic copies of traces may be readily obtained by pressing the slides in close contact with sensitized paper against a surface covered with black cloth, and placing for a moment or two in the direct rays of the sun.

The Manometric Tube.

An indispensable part of the instrument is the graduated glass tube. This serves as a diverticulum for the liquid, measuring the amount displaced and consequently the degree of expansion of the basal membrane. It shows the pulsations to the eye and indicates their amplitude and the pressure at which they are best developed. Thus it is, in a sense, a measurer of the arterial resistance, and as such is probably superior to any device yet employed for the purpose. Also it serves as a gauge for adjusting the tension of the basal membrane, and finally, by the indications of this tube the instrument is set in the best position for securing the best traces.

Uniformity of Action.

The instruments constructed alike, with the same chambers and bores, transmission tubes of the same calibre and length, index tubes of the same graduation, rubber membranes of the same quality and thickness, and these adjusted to the same tension and sensibility by means of devices provided, must yield uniform action and results. And this is so. Successive traces of the same pulse by the same instrument, are not more alike in form than traces of the same pulse by two instruments.

Multiplicity of Uses and Adaptations.

This instrument traces pulsations of the heart by placing the tube over the organ at point of strongest impulse, and then proceeding as in tracing the arterial pulse. It traces any accessible artery. It traces the pulsations of the fontanelle. It traces pulsating tumors.

It is converted into a *pneumograph* and traces the movements of respiration merely by confining the base to the chest wall by means of an inelastic belt round the thorax, the trace ascending in inspiration and descending in expiration; or by connecting the transmission tube with a flattened elastic tube surrounding the chest and filled with water, the line will be reversed, rising in expiration and falling in inspiration, as in Marey's pneumograph.

It is converted into a plethysmograph (*of Mosso*) and traces the totalized pulsation of the hand and forearm, simply by connecting its transmission tube with an apparatus of displacement containing these parts.

It traces muscular contractions and evidently is adapted to trace any and all the palpable movements of the organism. It is well adapted to experimentation on animals, and easily made to function as a *kymograph*. It performs well on the schema.

Tambour Polygraph.

If for any reason it should be deemed desirable to employ air instead of water as a transmission medium, the substitution may be made and without further change traces obtained, if the movements are not too feeble. Better results, however, are secured by employing a larger inscribing membrane. A tambour of suitable circumference is simply fixed over the chamber that was covered by the small membrane, and then with a vertical pin of proper length, connection is made with the lever and the apparatus is ready for work as a *tambour polygraph*.

Recording Surfaces.

Thin glass slides about five and three quarter inches long and width to suit, evenly smoked, are found, all things considered, the best form of writing surface for the great majority of work that one desires to do. But for long continuous lines of traces the lever may be turned upon a revolving cylinder covered with smoked paper, or upon a long band of paper made to advance by rollers, in the latter case the writing done with a suitable pen fixed on the lever and charged with ink. The instrument adapts itself readily to any form of recording surface found in physiological laboratories.

Multiple Simultaneous Inscriptions.

The crowning adaption is that by which the instrument admits of combination for the purpose of tracing two or more movements at the same time. The instruments are readily combined, in any number chosen, so as to write upon the same recording surface, but for convenience, and as answering the greater number of purposes, two are combined. These with accompanying chronograph and recording mechanism make a compact and convenient apparatus, that one is able to manage himself in clinical and much physiological work. Thus combined, the mechanism has many uses and brings to our knowledge numerous and varied important facts. A wide and promising field has been opened up by multiple simultaneous sphygmography. Marey's and the author's are the only mechanisms by which this method can be practised, and of their comparative merits we will speak soon.

Among the results already secured by this method are determinations of the time differences between the pulsation of the heart and different arterial pulses, and between the arterial pulses themselves, and consequently the velocity of the pulse-wave along the trunk and branches of the arterial tree, and the duration of the ventricular presphygmic interval; the variations of these time relations in health, and some abnormal variations in disease. The facts so gathered are valuable in themselves, and illustrate the kind of work that is easily done by the employment of a proper apparatus.

It is in multiple traces that sphygmography is able to show its strength. Single sphygmography has been in vogue since 1860, and although assiduously worked by skilful men, who have published their results, it has contributed comparatively little of real value; while the multiple method, only recently made practical, has already achieved a record of present utility and future promise. The one gives only a simple inscription of the form of the pulse; the other gives this, and besides gathers all the main facts pertaining to the movements of the heart and arteries, or other movements to which it is applied.

Explorators—Bases.

The devices for first receiving the movements are termed by Marey " *Explorateurs.*" They may more appropriately be called " bases." Any form of receiving base circumstances may call for readily adapts itself to the author's apparatus. The form shown in the cut has been settled upon, after many experimentings and provings, as the most eligible for clinical purposes, and at the same time as well answering for the greater part of physiological researches. A base of circular form answers well for receiving the movements of the heart, also of tumors, and the respiration, but after trial one finds in it no advantages outweighing that of having the two bases precisely alike. However, as just stated, for certain investigations bases of any form indicated may readily be attached.

Chronograph.

No registering apparatus for physiological and clinical work is complete without an accompanying chronograph, and, indeed, this is an indispensable attachment. The rate of progression of the recording surface is always an important element to be known and considered. The best-constructed movements are liable to variations of speed, yet it is easy to obtain a movement whose variation is so small as to be of little moment if the progression is measured. This the chronograph does, and notes every variation. The chronograph of this apparatus is neat and accurate, and fulfils its office with entire satisfaction. It is always running during experimentation, and traces its fifths of seconds on the lower part of the slide with every run of the same. Marey expresses himself much pleased with this principle, and as intending to add it to his new polygraph.

Convenience of Management.

The construction of the apparatus and the kind and variety of its work all go to set forth its special convenience. It is portable, easily kept in order, readily adjusted, and is applied with the greatest facility and without the least annoyance to the subject.

Comparison between Marey's New Polygraph and the Compound Sphygmograph.

Marey, seeing the advantages to accrue from such a combination, worked long to devise an apparatus that would trace the heart and pulse simultaneously on man, and finally produced his new polygraph, and with which he appears well satisfied. It is on the tambour principle, and consists of a *sphygmographe à transmission* and a cardiograph, filled with air and arranged to write on a revolving cylinder. With this apparatus the author has done very good work. Marey adopts the author's chronograph; in turn, his device for starting and stopping the movement, by displacement of air in a rubber tube held in the mouth, has been put to use. This adds much to the convenience of management, in that it leaves the hands free for manipulating the explorators, enabling one to successfully experiment on himself or others without an assistant.

It is manifest that the compound sphygmograph has certain advantages over the new polygraph, aside from the question of the superiority of water or air as a transmission medium. Its adaptability to trace a pulsation or movement in any situation, the uniformity and consequently reversibility of its bases, the showings and indications of its index tubes, and the facilities afforded for the graduation of its membranes to the same sensibility, are distinctive and advantageous features which pertain to the one and not to the other.

As to which is the better transmission medium, water or air, the comparative experiments which follow are in the line of elucidation.

Water and Air as Media of Transmission for Sphygmographs.

The utility of the graphic method as applied to clinical and physiological investigations is being demonstrated more and more. The success is to be accredited in large part to the employment of the method of simultaneous inscriptions. The sphygmograph and cardiograph, used separately, give permanent records of the arterial and cardiac pulsations, and show some features of form not otherwise distinctly revealed; but these instruments, combined in one so as to trace the heart and an artery at the same time on the same side, afford additional important indications, and such as are unattainable by any other means. Thus, two (or more) properly constructed inscribing instruments arranged in this manner are capable of showing all the palpable movements of the organism in their individual forms and phases, and the relations of these movements to each other. The most valuable of Marey's contributions in this department are the facts he has been able to bring to light by the method of simultaneous inscriptions.

A direct sphygmograph, as Marey's, although tracing correctly the form of the pulse, is not adapted to combination for the purpose named. To carry out the object requires that the movements be transmitted to a convenient distance where they may be recorded on the surface provided. To do this with fidelity and convenience has been the problem to be solved. Accordingly, Marey devised and has employed the tambour system, and utilized the contained air as the medium of transmission. The author devised and employed a modified tambour arrangement, and utilized water as the medium of transmission.

The pertinent question has arisen as to which of the two is the better medium. Marey, in his former experiments on animals, failed with water, but obtained brilliant results with air. In experiments on man, the author has succeeded with water, and found this medium so efficient and convenient that it left nothing to be desired. Marey believes that the inertia of the liquid column tends to deform the movements, while air, by its lightness, obviates this source of deformation. On the other hand, it is held that, in a proper apparatus, the inertia of the liquid column is reduced to an inappreciable minimum; and this medium, by its incompressibility and easy displacement, transmits the movements with unerring fidelity, while air is mistrusted on account of its easy compressibility. The question, however, can only be settled by the results of experiment.

The compound sphygmograph is constructed for water, yet it yields fair results when used with air. Whether these results will compare favorably with those of which Marey's improved polygraph is capable, there is at present no means of knowing; only among traces seen that were taken by the *sphygmographe à transmission*, which embodies the same principle, there are none superior to those the compound instrument is capable of, charged with air instead of water. To aid in the solution of the question the following comparative experiments were recently instituted.

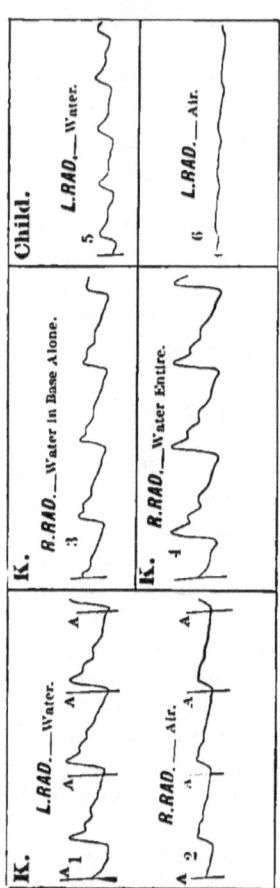

The author's two radial pulses are synchronous, as proven by many observations.

Experiment 1.—With the two instruments of the apparatus prepared and adjusted as usual, the liquid was removed from the one that writes below, and the other filled. Two assistants then applied them to the radials—the one filled with water to the left, the other filled with air to the right radial. They were manipulated until the levers displayed their best movements, and then the traces were taken. No. 1 shows the trace of the left radial by water; No. 2 of the right radial by air. This experiment was several times repeated with very similar results. There will be noticed the marked contrast in the amplitude and definition of the two traces.

Experiment 2.—The same basal chamber was completely filled with water, leaving the transmission tube and upper chamber containing air. The instrument was then applied to the radial and the best trace obtained; 3 is the result. This will be noticed as a very beautiful trace, of small yet larger amplitude than 2, and with secondary waves distinctly marked.

Experiment 3.—The same instrument completely filled with water and the right radial traced; 4 gives the result—a perfect trace, amplitude nearly twice that of 3 and about three times that of 2.

Experiment 4.—A child two and a half years old was the subject. With instrument adjusted as in the other experiments, 5 was obtained with water and 6 with air, as the best traces the same pulse would yield under the respective media.

These results all go to show that under the same or parallel conditions water produces a trace of much greater amplitude and better definition than air; although aside from difference of amplitude the two traces in general form are quite similar. The contrast proves that in transmission by water the force of the pulse is conserved, while by air it suffers loss; and herein lies the great superiority of water over air as a medium—water thus becoming applicable for tracing the weakest movements, while air in these conditions produces no tangible result. Both Marey and François Franck testify to the difficulty of obtaining traces of feeble movements by the *sphygmographe à transmission*.*

In regard to form, these traces † by water are as nearly faultless as any that could be produced by the best direct sphygmograph. It would be difficult to notice in them any effect of inertia of the liquid column. And No. 2 by air, so far as its meagerness will permit, shows a correct outline.

Attention is now called to traces 1 and 2, taken simultaneously. It will be observed that the position of the signal lines A, A, with respect to the beginning of the ascents, appears to be the same in both lines. If this be so, the beginnings of the traces are synchronous, and, the pulses being

---

* ". . .; malheureusement, le sphygmographe à transmission est moins sensible que l'instrument direct et ne donne de bons tracés que sur les sujets dont le pouls est assez fort."—"La Méthode Graphique," p. 585.

† Unhappily the engraver has failed in some instances, notably in the direction of the up-strokes, to follow exactly the lines of the originals.

synchronous, their beginnings must have been transmitted by the different media in equal times.

However, it should be remarked that a slight discrepancy may exist and be unnoticed, because of the uncertainty as to the exact position of the basal points in the long turns of the air traces. The result shows at least great promptness in the transmission by air, and whether this be really more rapid than by water will be seen farther on. The relationship shown is different from what had been, *a priori*, expected. It was inferred that the air under the stroke of the pulse would be compressed for a time, before the lever would rise, and thus the start would show delay as compared with that of water. But the fact would seem to be, that the air in order to actuate the instrument first becomes condensed under the requisite pressure, and then behaves as a practically incompressible medium within the forces engaged.

In order to test more rigidly the fidelity of the time results obtained with the two media as employed in the author's apparatus, the following further experiments were made:

Experiment 5.—Object: to test the correctness of water to show the time relation of two phenomena. *Details:* Apparatus, except the carriage, prepared and adjusted as usually employed in clinical observation. Tubes, twenty-six inches long between membranes, about one sixth of an inch bore, composed of alternate sections of glass and firm rubber tubing. Membranes respectively at standard tension; levers six inches long, with pin-socket half an inch from axis. The two bases were firmly secured in a block, side by side with membranes looking upward. Two similar corks transfixed by a metal rod were placed one upon each membrane, and a metal bar of sufficient weight was placed, with one end on the rod, the other on a support. A touch of the bar would instantaneously affect the basal membranes, and the motion in due time would be communicated to the levers, and written by them on the advancing slide. If the apparatus were true, the records of a motion thus communicated to the membranes would be synchro-

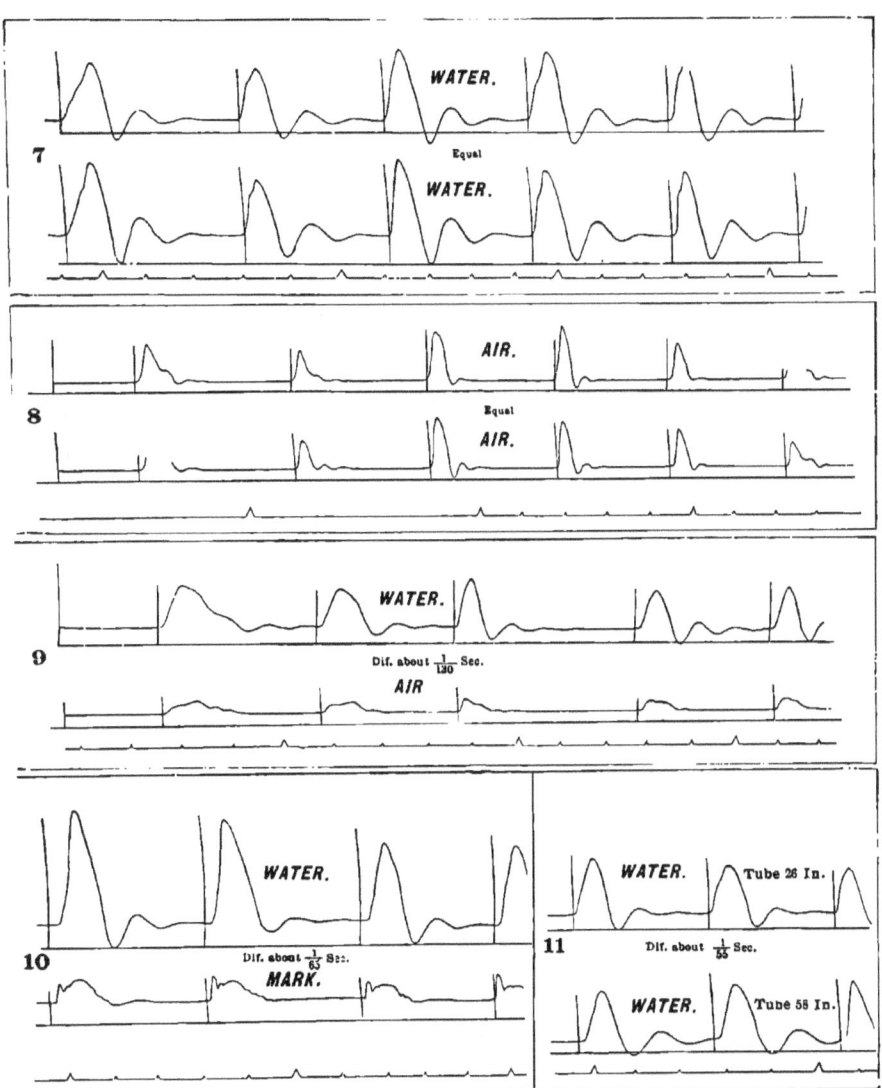

nous. To show more plainly any succession, a carriage was substituted whose speed was much greater than that ordinarily used. The usual chronograph, writing seconds and fifths, was employed. When all was ready, while the chronograph and slide were in motion, a series of quick taps were made with the finger on the bar, and thus were obtained two lines of traces. Immediately the slide was repassed, and halted at the points where the levers and basal points of the traces were in, or nearly in, conjunction, and then the levers were made to describe their curves by pressing on the membranes. The relation of these curves, termed signal lines, to the basal points, respectively indicates the synchronism or succession of the records. The result of the experiment is shown in No. 7. The basal points and signal lines, above and below, are precisely in the same relation, which proves that the two instruments transmitted and recorded the same movement in exactly equal times.

Experiment 6.—Object: to test the correctness of air to show the time relation of two phenomena. *Details:* The same as in last experiment, except that the instruments were charged with air instead of water. Result shown in No. 8. Here, too, the traces are uniformly synchronous.

Experiment 7.—Object: to test the difference in time of transmission by water and air. *Details:* The same as in Experiment 5, except that the upper instrument was charged with water and the lower with air. Result shown in No. 9. The eye can readily detect a difference in favor of an earlier elevation of the traces by air. The difference is small, and measures about $\frac{1}{140}$ of a second. The proof is positive that in this instrument a movement is more rapidly transmitted by air than by water.

Experiment 8.—Object: to determine the time consumed in the transmission of a movement by water through this instrument as usually employed. *Details:* A crank moving vertically on a central axle was so placed in relation to the instrument that one arm bore upon the basal membrane of the upper instrument, and the other arm supported directly the lever of the lower instrument. The same bar was placed with its end resting on the arm that pressed the membrane. Tapping the bar would raise the lower lever instantaneously and the upper after the movement had been transmitted through the channel of the instrument. Thus the experiment was made, and the result is shown in No. 10. The delay seems to measure about $\frac{1}{65}$ of a second.

Experiment 9.—Object: to determine the effect of tubes of different lengths on transmission by water through the author's instrument. *Details:* The same as in Experiment 5, except that the lower instrument was furnished with a tube fifty-eight inches long, while the tube of the upper remained twenty-six inches. Result shown in No. 11. Delay of the long tube over the other is measured at about $\frac{1}{55}$ of a second.

These experiments were all several times repeated, with great uniformity of the time results.

Conclusions.

1. Both water and air are reliable as media of transmission for movements within their respective ranges of availability.

2. Water transmits movements with much greater power than air, which gives it a wider range of application, embracing feeble movements that air is incapable of inscribing.

3. With a proper apparatus, properly adjusted, the form of a movement as transmitted by water is entirely correct; the form, also, as transmitted by air, may be correct.

4. The time relation of two movements may be uniformly correctly shown by either water or air—provided respecting the latter the movements be sufficiently strong.

5. The time lost in the transmission of a movement is greater by water than by air, and by either is in proportion to the length of the tube.

6. In order to show with precision the synchronism or succession of two movements, both instruments must be charged

wholly with water or wholly with air, and both transmission tubes must be of the same length.

In practical effect the foregoing experiments demonstrate the fidelity and availability of the water method to show the form and synchronism or succession of movements, great and small. They go, also, to show the fidelity of Marey's method for the purposes named, within the range of its availability. Hence, the results which have been achieved by the careful application of either method, may be accepted as rigidly true. If improvement in Marey's apparatus shall be found to adequately enhance the force of transmission, then the choice between the methods by air and by water will be governed by less weighty considerations than difference in energy of transmission.

Part I.
PHYSIOLOGICAL SECTION.

1

CHAPTER I.

FORM, RELATIONSHIP, AND CHRONOMETRY OF THE CARDIAC AND ARTERIAL MOVEMENTS.

The Velocity of the Pulse-Wave, and Duration of the Ventricular-Carotid and Ventricular-Presphygmic Interval in Young Children.

OBSERVATION 1.—Having removed a portion of the breast-bone of a living turtle, the base of the cardiograph was placed on the heart *in situ*, a finger touching the ventricle, to keep the organ in place under the instrument, and at the same time note the relationship of phenomena. Successively the heart felt as a hard ball and then as a flaccid sac. As the ventricle hardened, the spirit-index rose; as the ventricle softened, the spirit-index fell. The force turned upon the lever; it rose and fell, just as the liquid, with the hardening and softening of the ventricle. Tracing Fig. 7 was taken under these circumstances.

FIG. 7.—TRACING OF A TURTLE'S HEART.

Observation 2.—Severing the heart from its connections, it was placed upon a table, with its anterior aspect uppermost. It passed through its evolutions apparently the same as previous to its removal. It suddenly became hard to the touch, coincidently increasing its transverse diameter, bulging its anterior wall, twisting to the left, and tipping the apex forward at the end of the movement. Instantly following this stage, the ventricle became suddenly soft to the touch, coincidently extending its longitudinal diameter, letting fall its apex, flattening its anterior wall, and during these changes twisting toward the right. The cardiograph applied to the heart showed the upward sweep of the liquid column and writing-lever coincident with the first series, and the downward sweep of these indices coincident with the second series of phenomena described. Fig. 8 is the cardiogram traced under these circumstances.

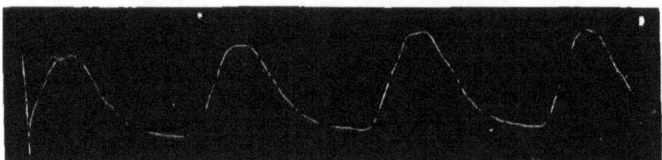

FIG. 8.—TRACING OF A TURTLE'S HEART AFTER REMOVAL FROM BODY.

Observation 3.—In man, the following coincidences and consequences of phenomena are determined: Marked impulse, as if a hard ball were thrust from within against the chest wall; elevation of the corresponding intercostal space or spaces; emission of the first sound of the heart; major rise of the spirit-index and writing-lever; these quickly succeeded by a lesser impulse; the second sound of the heart; depression of the intercostal space or spaces; and major fall of the spirit-index and writing-lever. Fig. 9 is the cardiogram of a healthy man.

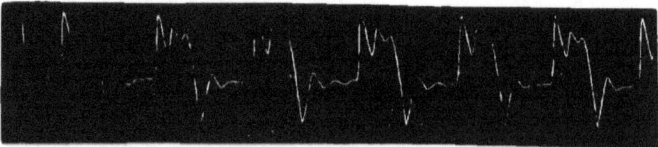

FIG. 9.—NORMAL HUMAN CARDIOGRAM.

Deductions.—Observations first and second together furnish positive proof that, in the tracing of a turtle's heart, the upward sweep is made by the systole, and the downward sweep by the diastole, of the ventricle. And the three observations together afford conclusive evidence that, in the normal cardiographic trace of man, the main up-stroke and the main down-stroke are produced respectively by the systole and diastole of the ventricles.

It has seemed important to be thus particular because of the evident fact that, until the relationship of the graphic lines to these two chief conditions of the heart's action is established, there can be no trustworthy interpretation of the cardiac and arterial pulse-curves.

In the prosecution of this work the apparatus used was that represented in Figs. 5 and 6 of previous pages. It consists essentially of two sphygmographs placed side by side, provided with one stage-movement for both instruments, and an *inscribing chronometer*, all arranged so that the three levers impinge at different heights upon the same slide. The time-writer, or *chronograph*, is a neat and correct instrument, and marks seconds and fifths with perfect accuracy.

With one base placed upon the heart, the other upon an artery, or each on a different artery, with the corresponding levers executing their best movements, and the chronograph going, when the clock-work that moved the carriage bearing the smoked-glass slide was started there resulted three graphic lines: one a cardiogram, one a sphygmogram—or two sphygmograms,—and one a *chronogram*. Tracings so taken with skill and care may be accepted as true and exact representations of the form, duration, and succession of the events under consideration.

A medical friend, aged thirty-two years, in good health, kindly submitted himself for experimentation, and aided with his intelligent and skilful assistance. When the cardiac and arterial tracings were being taken together, the subject occupied the left prono-lateral position, and held the cardiac base in place himself, whilst the author held the arterial base and manipulated the apparatus, as he did likewise when the pairs of arterial tracings were taken, the subject also holding one base *in situ*. The stage-movement was set at a rather

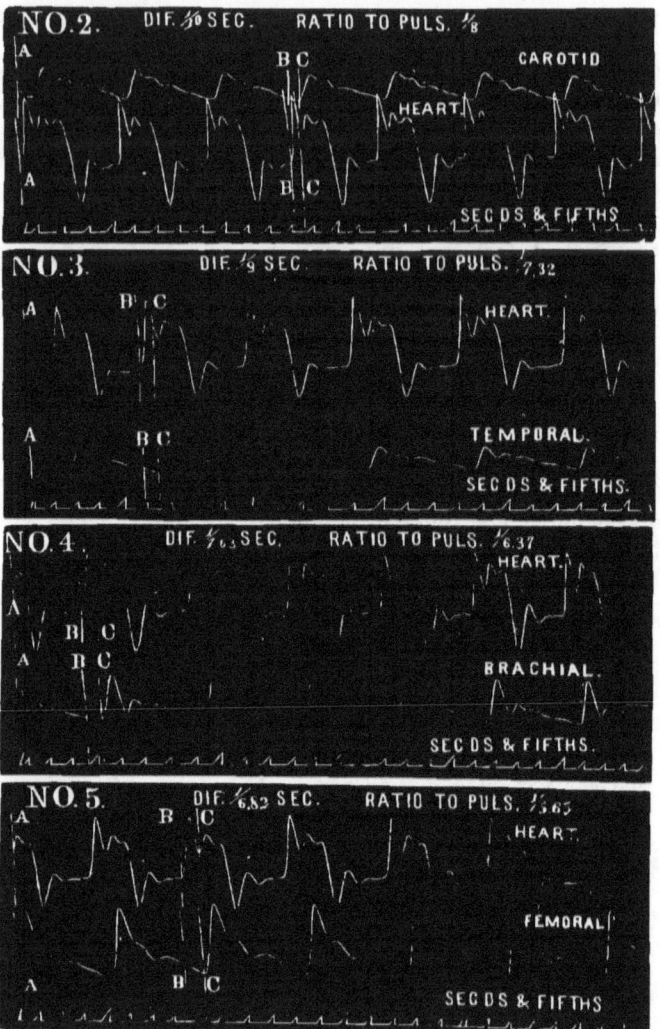

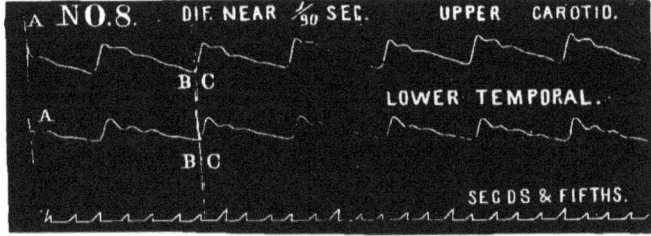

greater speed than usual, in order to show better the curves and distances. The adjustments were all in order, and every care was observed to insure truthful results.

The tracings as produced were marked with the name of the artery, or arteries, and then fixed by the accustomed method. In due time they were prepared for the engraver by adding the lines and inscriptions which appear in the reproduction, and are as follows: A, of each tracing, is the curved line already on the glass, made by the lever before the start of the carriage, and from some point in which line the tracings begin. The lines B and C were drawn with a compass whose span equalled the length of the levers; the glass and fixed point of the compass being in the same relative position as the glass and pinions of the levers respectively, so that the lines described by the free point of the compass would be parallel throughout with lines described by the levers on the glass at rest. In this manner B was drawn through the point of ascent of the cardiac trace, and C through the point of ascent of the arterial trace. Or, where two arteries were under experiment, B was drawn through the point of ascent of the proximal, and C through that of the distal. The curve B, on the pulse-line, was drawn at the same distance from A of this line as that between A and B of the heart-line; and the curve C on the heart-line was drawn at the same distance from A of this line as that between A and C of the pulse-line. The same rule was observed in drawing the curves for the pairs of arterial traces. The distance $B\ C$ measures in horizontal line the difference between the points of ascent of the cardiac and arterial traces represented. The time value of $B\ C$ is ascertained by its measurement on the proper part of the chronogram, and to indicate which the lines A and B from the lower tracing were carried down. The other markings need no special designation here, but will be referred to in the proper place. The engravings are faithful reproductions.

It is evident that the series represents the form, relation, duration, and succession of the movements pertaining to the same pulsations, as they emanate from the heart and display themselves in different arteries. Both in cardiac and arterial traces the space between the successive basal points of the main ascending lines measures an entire round of pulsation, and, in the corresponding pulsations of the respective pairs, these distances are found to be equal. The distance $B\ C$ running through the series, and seen of different lengths, represents, as before remarked, the *asynchronism* between the beginning of these corresponding pulsations.

The chronometric differences, obtained by the means and in the manner above described are as follows:

No. 1 presents a tracing of the heart, the carotid artery, and the divisions of time. $A\ B$ shows the distance to the commencing cardiac pulsation, and $A\ C$ the distance to the commencing carotid pulsation; while $B\ C$ shows the difference between these beginnings. $B\ C$, taken accurately between the points of hair dividers, and then placed on a hundredth-inch scale, under a magnifying glass, was found to span $6\frac{3}{4}$ of the divisions, while the fifth of the second, of which $A\ B$ is a part, was found to span $13\frac{1}{2}$ of the divisions. So the formula stands: $A\ B = \frac{6\ 3\cdot4}{13\ 1\cdot2}$ of $\frac{1}{5} = \frac{1}{10}$ second. Ratio to plusation, $\frac{1}{6}$.

No. 2 is a repetition of No. 1, the tracings taken with the bases reversed to show that the difference obtained was not produced in the instrument, but by the action of the organs on which it was placed. The difference, worked out as in the preceding, gives the formula: $B\ C = \frac{6\ 1\cdot2}{13}$ of $\frac{1}{5} = \frac{1}{10}$ second.

No. 3 contains a tracing of the heart, and temporal artery in front of the ear, with accompanying chronogram. The difference is expressed by: $B\ C = \frac{8}{14\ 1\cdot2}$ of $\frac{1}{5} = \frac{1}{9.00}$ second. Ration to pulsation, $\frac{1}{7.32}$.

No. 4 shows the pulsations of the heart, and brachial artery at the bend of the elbow; also the time. Difference expressed by: $B\ C = \frac{9\ 1\cdot2}{14\ 1\cdot2}$ of $\frac{1}{5} = \frac{1}{7.63}$ second. Ratio to pulsation, $\frac{1}{6.37}$.

No. 5 is a tracing of the heart, and femoral artery at the groin, and the time. Formula of difference expressed by: $B\ C = \frac{11}{15}$ of $\frac{1}{5} = \frac{1}{6.81}$ second. Ratio to pulsation, $\frac{1}{5.63}$.

No. 6 is of the heart and radial artery, with time-line. Formula of difference: $BC = \frac{10}{13}$ of $\frac{1}{5} = \frac{1}{65}$ second. Ratio to pulsation, $\frac{1}{56}$.

No. 7 shows the pulsations of the heart, and posterior tibial behind the malleolus, with corresponding time. Formula of difference: $BC = \frac{15}{12\cdot 15}$ of $\frac{1}{5} = \frac{1}{4\cdot 10}$ second. Ratio to pulsation, $\frac{1}{3\cdot 33}$.

No. 8 shows tracings of the carotid, and temporal in front of the ear, and the time. Formula of difference: $BC = \frac{14}{15}$ of $\frac{1}{5} = \frac{1}{50}$ second.

No. 9 shows tracings of the femoral at the groin, and posterior tibial behind the malleolus, with the time. Formula of difference: $BC = \frac{2}{14}$ of $\frac{1}{5} = \frac{1}{10}$ second.

The original glasses give perfect delineations of these asynchronisms. The measurements were made from the glasses. For the figures given a limited margin is asked on account of the difficulty of defining precisely the point of ascent; this, especially in the arterial traces, being more or less obscured in a curve. It is believed the fractions given do not err in any instance to exceed the one hundredth of a second. Many tracings have been taken besides these now recorded, and the differences calculated from them, and, while the results vary somewhat under modifications of the circulation consistent with health, the average for a healthy man corresponds with the figures announced.

It will be noticed that the *fifths* of seconds vary in the measurement from twelve and a half to fifteen hundredths of an inch. At first view this might be taken as showing irregularity in the time measurement; but, in reality, it notes the unequal speed of the carriage. As more or less inequality pertains to the run of the best-constructed stage-movements, the necessity of the chronograph becomes apparent.

A second set of tracings are presented in two series: one from a man, in health, aged fifty years; the other from a man, in health, aged twenty-five years. The radial pulse of the older subject, designated by K, is usually 72 to 76 per minute, but runs faster under experimentation; it is regular, ample, and resisting to the fingers. In the tube, the highest undulations are displayed at 16° of pressure, the same rising and falling one and a half to two degrees; decline gradual, and marked by the lesser oscillations. The form of the tracings, as seen, confirms these indications, and demonstrates the pulse to be *above* the average in tension and resistance to pressure.

The radial pulse of the younger subject, designated by L, runs from 62 to 72 per minute; it is ample, bounding, dicrotous, and compressible to the fingers. In the tube, the undulations are highest at 12° of pressure, rising and falling two to two and a half degrees. The form of the tracings, as seen, confirms these indications, and demonstrates the pulse to be *below* the average in tension and resistance to pressure.

The advantage, thus, of two sets of tracings, which, while proving each other, and showing the relations sought to be determined, serve to demonstrate very closely the physiological chronometric range and average, in the movements of the adult heart and arteries, becomes forcibly manifest.

Demonstrations.

Problem 1.—To determine the average time-difference between the carotid and dorsalis-pedis or posterior-tibial pulse.

Solution by Plates No. 11 of the K series and No. 17 of the L series. K's time, $0''.125 = \frac{1}{8}$, and L's time, $0''.166 + = \frac{1}{6}$ of a second. Mean result, $0''.1458 = \frac{1}{6.85}$ of a second.

Problem 2.—To determine the average time-difference between the carotid and femoral pulse.

Solution by No. 12 of the K and No. 18 of the L series. K's time, $0''.05 = \frac{1}{20}$, and L's time, $0''.0909 = \frac{1}{11}$ of a second. Mean result, $0''.0704 = \frac{1}{14.2}$ of a second.

Problem 3.—To determine the average time-difference between the femoral and dorsalis-pedis or posterior-tibial pulse.

Solution by No. 13 of the K and No. 19 of the L series.

K's time, $0''.075 = \frac{1}{13.33}$, and L's time, $0''.0714 = \frac{1}{14}$ of a second. Mean result, $0''.0732 = \frac{1}{13.66}$ of a second.

Problem 4.—To determine the average time-difference between the carotid and radial pulse.

Solution by No. 14 of the K and No. 20 of the L series. K's time, $0''.0714 + = \frac{1}{14}$, and L's time, $0''.088 + = \frac{1}{11.35}$ of a second. Mean result, $0''.0797 = \frac{1}{13.54}$ of a second.

Problem 5.—To determine the average time-difference between the radial and dorsalis-pedis or posterior-tibial pulse.

Solution by deducting K's carotid-radial time, No. 14, from his carotid-dorsal, No. 11, which gives $0''.0536 = \frac{1}{18.66}$ of a second, and by No. 21 of the L series, which gives $0''.0625 = \frac{1}{16}$ of a second. Mean result, $0''.058 = \frac{1}{17.33}$ of a second.

Problem 6.—To determine the time relation between the femoral and radial pulse.

Solution by No. 15 of the K series and No. 22 of the L series. In K the femoral precedes the radial by about $0''.02 = \frac{1}{50}$ of a second, while in L the radial precedes the femoral by a time too short for any thing like accurate measurement. Mean result, antecedence of the radial pulse probably not longer than $0''.01 = \frac{1}{100}$ of a second.

Problem 7.—To determine the average time-difference between the systole of the ventricle and the carotid pulse.

Solution by No. 16 of the K and No. 23 of the L series. K's time, $0''.077 = \frac{1}{13}$ of a second; L's time, $0''.100 = \frac{1}{10}$ of a second. Mean result, $0''.0884 + = \frac{1}{11.33}$ of a second.

From the above facts are deduced the following:

Corollary 1.—In different individuals the time difference of the pulse between the same designated arterial points is subject to marked inequality.

Corollary 2.—In such comparison the asynchronism between the carotid and femoral pulse shows the greatest diversity.

Also, from the above, and other proper data of the plates, is deduced:

CARDIAC AND ARTERIAL MOVEMENTS.

Corollary 3.—In the same individual the time difference of the pulse between the same designated arterial points, as noted at different times, and even in successive pulsations, is liable to a limited variation.

The next problems concern the rate of transmission of the pulse-wave as a whole along the arterial lines. The solutions, in addition to the data already presented, call for those expressing the arterial distances between the points under observation. The latter have been approximately ascertained by careful external measurements, and will be stated in order as required.

Problem 8.—To determine the average *mean* velocity of the pulse-wave along the arterial tree from the trunk near the root to a branch in the foot.

Solution.—K measures from the third cartilage point, opposite the aortic orifice, to the carotid and dorsal points respectively, 7 and 53 inches. Six inches is added to the latter distance, to cover the aortic arch. Evidently, then, the distance represented by the difference between the carotid and dorsalis-pedis pulse is: $53 + 6 - 7 = 52$ inches.

Carotid-dorsal time difference, $0''.125$, is to distance traversed by pulse-wave (52 inches) as $1''$. is to the velocity of pulse-wave per second, viz., 416 inches. The same operation applied to L yields the formula: $0''.166 + : 51$ inches :: $1''$. is to the required answer, viz., $306 +$ inches per second. Mean velocity, 361 inches per second.

Problem 9.—To determine the average velocity of the pulse-wave along the aorta and iliacs to the femoral at the groin.

Solution.—K's carotid-femoral difference or transit time (No. 12), $0''.05$; distance traversed, 17 inches. Result, 340 inches per second. L's carotid-femoral transit time (No. 18), $0''.909$; distance traversed, 18 inches. Result, 198 inches per second. Mean velocity, 269 inches per second.

Problem 10.—To determine the average velocity of the pulse-wave along the arteries of the inferior extremity from the femoral at the groin to the dorsal of the foot.

Solution.—*K*'s femoral-dorsal transit time (No. 13), 0".075; distance traversed, 35 inches. Result, 466 inches per second. *L*'s femoral post-tibial transit time (No. 19), 0".0714; distance traversed, 33 inches. Result, 462 inches per second. Mean velocity, 464 inches per second.

Problem 11.—To determine the average velocity of the pulse-wave along the arteries of the upper extremity from the subclavian, at a point seven inches from the heart, to the radial at the wrist.

Solution.—*K*'s carotid-radial transit time (No. 14), 0".0714; distance traversed, 23 inches. Result, 322 inches per second. *L*'s carotid-radial transit time (No. 20), 0".088 +; distance traversed, 23 inches. Result, 258 inches per second. Mean velocity, 290 inches per second.

Data are now at command for solution of—

Problem 12.—To determine the average duration of the pre-sphygmic portion of ventricular systole; or, in other words, the interval between the beginning of ventricular contraction and that of aortic expansion.

Solution by Calculation.—The velocity of the pulse-wave between the ventricle and carotid must be essentially the same as that along the aorta and iliacs. The measurement between the ventricle and carotid point is seven inches. Hence, by these data, K's transit time over the distance between the heart and carotid point is 0".0206, which, deducted from K's time difference between these points, viz., 0".077, gives the result, 0".0564. The same operation carried through L gives the result, 0".647. Therefore the mean result is 0".605 = $\frac{1}{16.53}$ of a second.

Solution by Direct Demonstration.—In K, No. 16, line 2 cuts the cardiac trace at the point which marks the end of systole, and line 2' cuts the carotid trace at the apex of the second wave, which also answers in the pulse to the cessation of cardiac systole. B and C, as usual, mark respectively the beginning of ventricular contraction and that of arterial expansion. Then the space B 2 represents the whole of ventricular systole, and C 2' the whole of pulse-expansion due directly to ventricular systole; and the difference between these distances represents the duration of the pre-sphygmic portion of ventricular systole. $C\,2'$ placed within $B\,2$ spans from 2 to the dotted line. Hence, the space between B and the dotted line is the interval sought. This measures by the chronogram $0''.0564 = \frac{1}{17.73}$ of a second. The same process applied to L, No. 23, yields the result, $0''.0647 = \frac{1}{15.45}$ of a second. Mean result, $0''.0605 = \frac{1}{16.52}$ of a second.

The subjoined table affords a compact record of the leading facts so far demonstrated:

POINTS UNDER EXPERIMENT.	Arterial Distances traversed by the Pulse-Wave.	Mean Time-Differences of Pulse-Wave between the Points designated.	Mean Velocity per Second of Pulse-Wave along the Arteries included.
	Inches.	Seconds.	Inches.
Carotid and dorsalis pedis	52	$0''.1458 = \frac{1}{6.85}$	361
Carotid and femoral	17 and 18	$0''.0704 = \frac{1}{14.2}$	269
Femoral and dorsalis pedis	35	$0''.0732 = \frac{1}{13.66}$	464
Carotid and radial	23	$0''.0797 = \frac{1}{12.54}$	290
Heart and carotid	7	Mean time-difference between pulsations, $0''.884 = \frac{1}{1.13}$ sec. Mean transit-time from aortic orifice to carotid point, $0''.0279 = \frac{1}{35.84}$ sec. Mean ventricular pre-sphygmic time, $0''.0605 = \frac{1}{16.52}$ sec.	

From the foregoing demonstrations and data of the cuts, the following corollaries are deduced:

1. The rate of transmission of the pulse-wave along different portions of the arterial tree is not uniform, but considerably diverse.

2. The rate is minimum for the aorta, maximum for the arteries of the lower extremity, and intermediate for those of the upper extremity.

3. Along the same arterial line the rate increases as the distance from the heart increases.

4. In the same healthy individual, in the same arteries, the rate is subject to a limited variation.

5. In different healthy individuals, in the same arteries, the rate is subject to marked diversity, of which the widest is in the aorta.

6. Both in the same and different healthy individuals, the pre-sphygmic portion of the systole of the ventricle is liable to considerable variation.

Explanation of Plates.

Tracings are in pairs, one above the other, taken at the same time, on the same glass. On most of the plates the same arteries are traced by reversal of the bases, for the purpose of proof. Near the lower margin of each plate is the time-line, showing fifths of a second between the points. The arteries will be recognized by the abbreviation near the trace. A, A', are the curved lines made by the levers before the start of the carriage; and a is made in like manner by the time-tracer, and marks the beginning of the chronogram. B is a line drawn, parallel with A, through the basal point of the proximal trace; and C is a line drawn, parallel with A', through the basal point of the distal trace. The space $B\,C$ is the difference between the proximal and distal pulsation, and figures express the value of $B\,C$ in fractions of a second, as carefully computed from the chronogram.

1, 2, 3, are lines parallel with A and B, cutting the apices of the first and second waves, and the aortic notch of the third or aortic wave, respectively; and $1'$, $2'$, $3'$, are lines parallel with A' and C, cutting the distal pulsation at the same respective distances from C as 1, 2, 3, are from B.

The figures within a pulsation indicate its frequency per minute.

On Nos. 16 and 23 the dotted line between B and C divides $B\,C$ into pre-sphygmic and transit time. PR. is for pressure, and the following figure indicates the number of degrees, by the tube, at which the traces were taken.

Int., interval; dist., distance; vel., velocity. The other abbreviations cannot be mistaken.

The fine line leading from $B\,C$ to the time-line shows the proper space from which the difference was estimated.

Problem 13.—To determine, in the arteries under observation, the *rule* of the time relation of the three principal sec-

The K″ Series, from a Man, in Health, aged 50 Years.

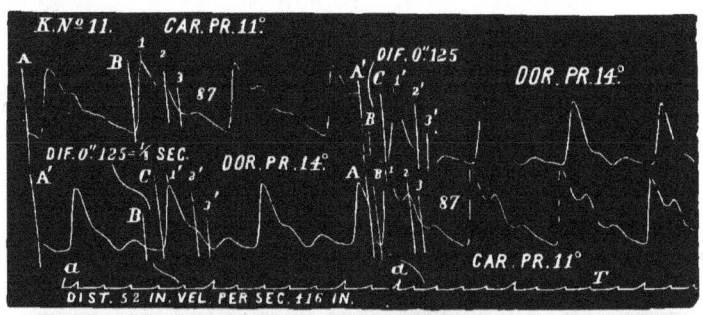

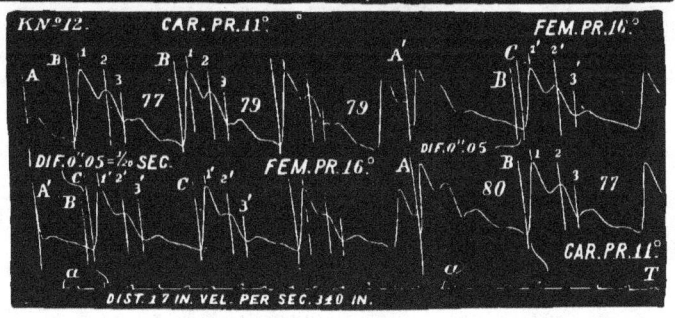

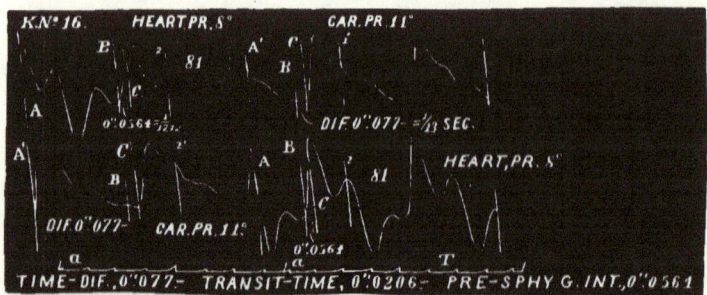

The L Series, from a Man, in Health, aged 25 Years.

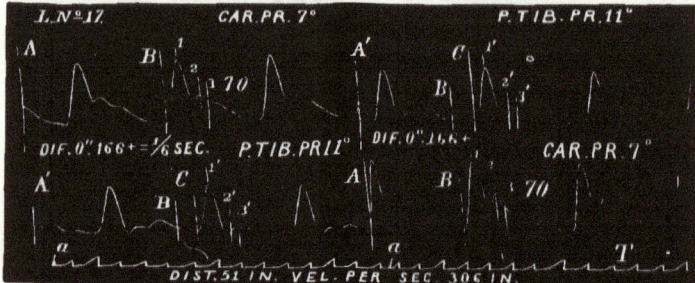

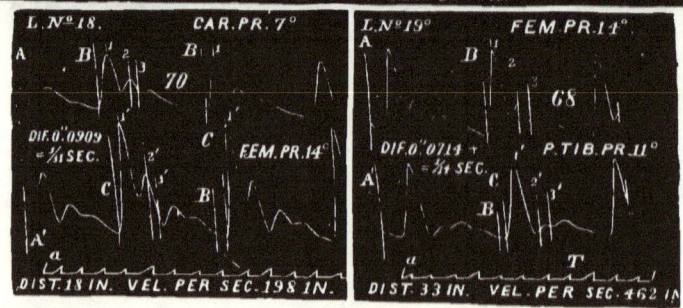

THE L SERIES.—(Continued.)

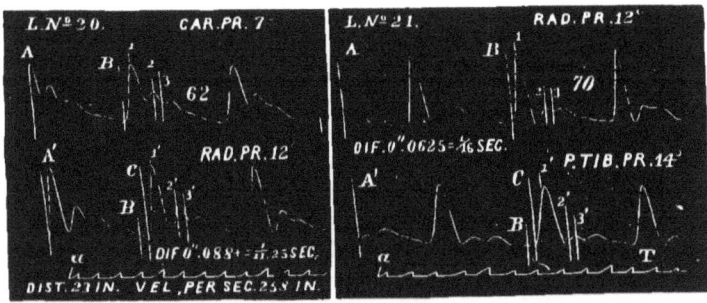

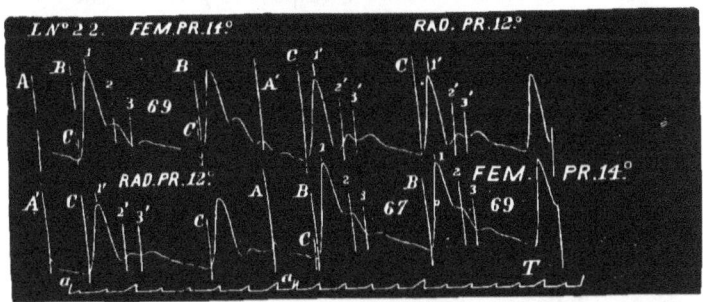

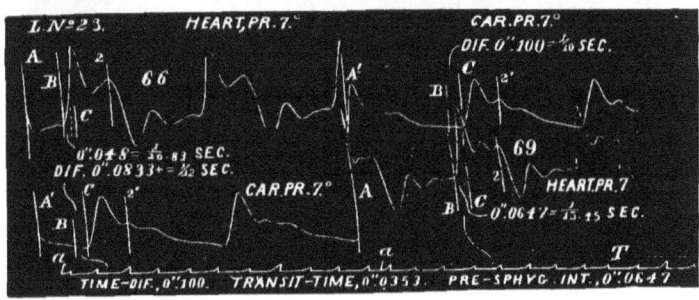

ondary waves to the beginning of the pulsation of which they are parts.

While fully appreciating the difficulty of an exact rendering of the specific facts embraced in the *chronometry* of the secondary waves, the data in hand are offered as competent for the solution of the problem as stated above.

Solution, by a critical examination of the plates. The position of the lines 1', 2', 3', on the distal trace, as to the summits of the first and second waves, and the aortic notch or beginning of the third wave respectively, shows the relation sought; inasmuch as these numbered lines are placed at the same respective distances from *C* as 1, 2, 3, on the proximal trace, and exactly cutting the secondary waves, are placed from *B*.

In the reading it is important to consider that the aortic is truly a double wave, and that the first indentation is properly the aortic notch. The double form is shown distinctly in the carotid, and is more or less indicated in the femoral traces. In the radial, dorsal, and posterior tibial, it scarcely appears.

Examination of the pairs of the *K* series, all shown by reversal, yields the following results:

1. The first and second waves of all the pairs attain their summits as early in the distal as in the proximal arteries—in the dorsalis-pedis as in the carotid pulse.

2. The beginning of the third or aortic wave is slightly but clearly delayed in the dorsal as compared with the femoral and carotid pulse, while delay is scarcely shown in the femoral and radial as compared with the carotid pulse, and in the femoral-radial pair the correspondence is complete.

Examination of the pairs of the *L* series (part shown by reversal) yields the following results:

1. The first wave is shown to attain its summit at the same time in the posterior-tibial and carotid pulse (shown by reversal); in the femoral and carotid; in the radial and carotid; in the femoral and radial; while the summit is shown

CARDIAC AND ARTERIAL MOVEMENTS. 47

very slightly delayed in the posterior tibial as compared both with the femoral and radial.

2. The second wave is shown slightly antecedent in the femoral and subsequent in the radial, as compared with the carotid ; delayed in the radial as compared with the femoral, and in the posterior tibial as compared with the femoral, radial, and carotid.

3. The beginning of the aortic wave is shown delayed in the distal pulse of all the pairs except the femoral-radial, in which it marks corresponding time.

Allowing for fallacies, and giving the above results a judicious interpretation, the premises will justify at least the following statement:

Rule 1.—The interval between the beginning of the pulse and its acme of expansion is the same in all parts of the arterial system.

Rule 2.—In certain conditions of the vessels and circulation the second wave keeps close time with the first in the onward flight, while in certain other conditions of the same the second falls notably behind the first in the progress from the heart.

Rule 3.—The aortic wave rises later in the distal than in the proximal pulses, and latest in the pulse most distant from the heart.

Are the foregoing data reliable? With certain reservations and restrictions, they are. The time differences shown on the engravings are not the certain expressions of the exact asynchronisms between the pulsations represented, but they are, indeed, extremely fine approximations thereto. The basal point of ascent from which the measurements were made is notably the most stable of any in a tracing. In the mechanism the lever has descended and become poised for the moment before it mounts up again on the current wave. Thus the basal point is wholly removed from the disturbing influence of inertia of the lever. The other chief obstacle to a good tracing—namely, undue friction of the writing-point against the slide—is so palpable when present, and so easily obviated, that this cause of displacement of the basal point need never be operative. Indeed, so little liable is this point to fallacious deviation, that its indications may be relied upon even in tracings not altogether faultless in form. In the tracings given, the basal point of each and every pulsation is unquestionably in its true position. The instrument, properly charged and adjusted, and used with skill and care, is simply incapable of erroneous registry of this point. Whence come, then, the errors admitted as liable in the representations of the time differences? The automatic registry is perfect, but the estimation thereof is imperfect.

The measurements are subject to fallacy from two sources: one, the difficulty of exactly locating the basal point, more or less obscured as it is in a curve; the other, an unnoted change in the speed of the carriage taking place within the limits of the time-points from which the measurements are made. The first source, it would seem, is unavoidable, and must be continuous; the second may be obviated when we attain to a mechanism that will move the slide with a certain unvarying speed. As no error attaches to the time-line, and the time differences were carefully computed from it, deviations on account of unequal movement of the slide are at a minimum; and, indeed, aberrations from the causes named, even when acting in conjunction, are too inconsiderable to affect appreciably the value of the data determined.

The method by reversal of the bases obviously affords positive proof that the instrument was delicate and true, and gave the correct difference between the points to which the bases were applied.

Each pair of tracings presented was selected from many taken from the points designated, the asynchronisms in all having been measured and noted. The small range of variation in the measurements was to an extent equalized in the representation by choosing and marking a pulsation whose time difference, as ascertained, was a near average of the observations. And yet these individual variations, while real, come in to mar the harmony of results, and prevent agreement in figures it would be satisfactory to have. Thus, the carotid-post.-tibial time should be equal to the sum of the carotid-femoral and the femoral-post.-tibial time, and either of these latter subtracted from the first should leave exactly the other. This agreement does not quite obtain in the L series. However, these discrepancies are small, and not serious.

In the tracings by reversal, pulsations were selected—one pair from each other—whose time differences were equal.

In L's cardiac-carotid tracing, No. 23, two time differences of unequal value are marked—one on each side of the reversal lines. This was done to show the variation which may be noted in the time difference between the heart and carotid in so short an interval. The estimates were made from the longer time, because this is L's more usual time between these points.

The measurements of the arterial lengths included between the arterial points are approximations, but must be so near the true distances that but very small error can arise from this source in the calculation of wave-velocities.

In regard to the data for determining the relative chronometry of the secondary waves, it is proper to premise that such data can only have value in the best-formed tracings. Between the basal point and aortic notch, friction and inertia exert their disturbing sway, and in consequence the apices of the first and second waves are frequently traced out of their true position. The aortic notch is less influenced than the preceding waves by extraneous causes, and, next to the basal point, it is the most stable. The summit of the aortic wave is uncertain.

The tracings given are free from distorting effects of friction, for all were taken with as light pressure of the tracer as possible to secure delineation. Inertia of the lever does not seem to have had appreciable effect upon the K series, and the reversal showing the points of the waves in the same relative position proves the perfection of this record.

In a part of the L series effects of inertia seem visible. This would be expected in a pulse of low tension and high amplitude. Supposed fallacies are : *a*, postponement of the apex of the first wave of the femoral and of the radial, compared each with the posterior tibial ; *b*, *antecedence* of the second wave of the femoral compared with the carotid ; *c*, excessive postponement of the second wave of the radial compared with the femoral ; and *d*, excessive postponement of the second wave of the posterior tibial compared with the radial and femoral. In other respects, the *L* series would appear to be a true exposition of the relations of the secondary waves.

The relations shown of the femoral and radial pulse to each other, in the two series respectively, afford striking confirmation of the fidelity of the entire exposition. The femoral pulse in the *K* series, notably *preceding* the radial, is what must be if the expressed time-differences and velocities between the carotid-femoral and carotid-radial are correct ; while the femoral pulse in the *L* series, slightly *succeeding or about equaling in time* the radial, is what must be if the expressed time-differences and velocities between the carotid-femoral and carotid-radial are correct. The contrast has been noted in at least a score of tracings from these subjects, and in no instance has it failed to be observed.

In the preparation of the glasses, the added lines were drawn with exceeding care, and the proper figures and letters written in their places. The transfer to wood was effected without change, by photography—the glasses used as negatives in direct contact with the sensitized blocks; skilful cutting completed the work. The reproduction appears perfect.

The Velocity of the Pulse-Wave and Duration of the Ventricular-Carotid and Ventricular-Presphygmic Interval in Young Children.

By the method already detailed and with equal care a considerable number of observations have been made on the velocity of the pulse-wave in young children ; and the results of these are fairly represented in the subjoined series of tracings procured from a healthy little boy aged 4½ years. The tracings were all taken at one sitting; and, although defects of form are plainly visible, they are perfect as regards the basal points of ascent and representation of the time differences which alone concern the immediate investigation. Finer-formed tracings could have been obtained by repeating the experiments, and multiplying the sittings, but this would have consumed unnecessary time and enforced upon the little subject undesirable restraint. The irregularity of rhythm is not greater than that which is often observed in the pulse of young children.

The temporal in front of the ear was chosen instead of the carotid, because the former answered the purpose very nearly as well, and the latter could not be subjected to sufficient pressure without annoyance to the child.

The space between the lines B and C, as heretofore, shows the time difference, and the figures express the same in decimals of a second, as computed from the chronogram marking fifths of a second between the points.

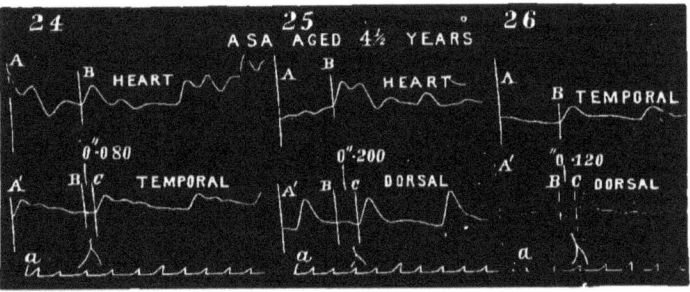

52   THE SPHYGMOGRAPH.

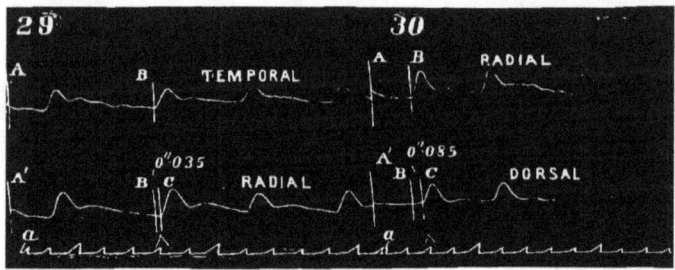

Below is a table of the child's measurements and data, together with a tabular restatement of *L*'s and *K*'s data for the purpose of more ready comparison.

ASA'S MEASUREMENTS.

Heart to dorsal, 31 inches + 4 in. for aortic arch =	35 in
Heart to femoral, 11 " + 4 " " =	15 in
Heart to radial . . .	18 in
Heart to temporal . . .	9 in

ASA'S DATA.

	Time.	Velocity per Sec.
Temporal-dorsal distance 35 — 9 = 26 in.	.120 sec.	216.6
Temporal-femoral " 15 — 9 = 6 in.	.025 "	240
Femoral-dorsal " 20 in.	.095 "	210
Temporal-radial " 18 — 9 = 9 in.	.035 "	257
Radial-dorsal " 17 in.	.085 "	
Heart-temporal " 9 in.	.080 "	
Heart-dorsal " 35 in.	.200 "	
Computed presphygmic interval, .080 — .035 = .045		

L'S DATA.

	Distance.	Time.	Velocity per Sec.
Carotid-post.-tibial	51 inches.	.166 sec.	306 in.
Carotid-femoral	18 "	.0909 "	198 in.
Femoral-post.-tibial	33 "	.0714 "	462 in.
Carotid-radial .	23 "	.0888 "	259 in.
Radial-post.-tibial		.0625 "	
Heart-carotid .	7 "	.1000 "	
Presphygmic interval .		.0647 "	

K'S DATA.

	Distance.	Time.	Velocity per Sec.
Carotid-dorsal	52 inches.	.125 sec.	416 in.
Carotid-femoral	17 "	.050 "	340 in.
Femoral-dorsal	35 "	.075 "	466 in.
Carotid-radial	23 "	.0714 "	322 in.
Radial-dorsal		.0536 "	
Heart-carotid	7 "	.077 "	
Presphygmic interval		.0564 "	

Evidently the wider the space $B\ C$, and the wider the distance between the points under experiment, the closer may be the estimated time-differences and velocities; while the significance of small errors in the estimates increases as these spaces diminish. Accordingly in Asa's case, while the estimated time-difference and velocity of the pulse-wave between the temporal and dorsal cannot err materially, the estimates of the divisions may err so as to mar in some degree the figures of the apportionments. However, the admissible errors by no possible just construction can be made sufficient to annul the marked contrast in the apportionments as shown between the child and adult.

A review and comparison of the data as set forth demonstrate the following propositions:

1. *The mean velocity of the pulse-wave in the arterial tree is much slower in young children than in adults.*

2. *In such comparison, the greatest diversity is in the lower extremities, where the velocity of the pulse-wave in young children may not exceed one half that in adults.*

3. *While in adults the velocity of the pulse-wave is much faster in the lower extremities than in the trunk and upper extremities, in young children such difference does not obtain.*

Also the data indicate almost to a demonstration that proposition 3 might be stated thus: *While in adults the velocity of the pulse-wave is much faster in the lower extremities than in the trunk and upper extremities, in young children the order may be reversed, though with a nearer approach to equality between the lower and upper divisions.*

Respecting the comparison of velocities in the trunk and upper extremity, in young children, as shown by Asas' figures, these are so close as to preclude a positive conclusion as to which preponderates. The same is the case in K, while in L a notable preponderance is shown in favor of the upper extremity. In the trunk, it is shown that Asa's velocity is greater than L's, and much less than K's. In the upper extremity, it is shown that Asa's velocity is about equal to L's, and considerably less than K's.

From a consideration of all the showings, the inference seems just, as a general proposition, *that the mean velocity of the pulse-wave increases with increase of age.*

In young children, closely associated with the pulse-wave velocity are: (*a*) the time difference between the heart and a near available artery, and (*b*) the duration of the ventricular presphygmic interval. In the investigations attention also has been given to these points. No. 24 of the series gives a fair representation of the results as to the entire difference in time between the cardiac and temporal pulsations. The showing is plain, and the measurement is placed at .08 of a second. The presphygmic interval is arrived at by deducting the transit time, as computed, on the basis of the wave-velocity in the upper extremity, at .035 of a second, from the entire time-difference. Thus · .080 — .035 = .045 of a second.

So from the data we are enabled to formulate two interesting facts:

1. *The time difference between the beat of the heart and the carotid pulse is very nearly the same in young children as in adults.*

2. *The interval between the contraction of the ventricle and expansion of the artery is notably less in young children than in adults.*

The reason of the phenomena we are becoming acquainted with will be an interesting study; but at present we are concerned with the immediate facts, well knowing that the explanations will be easier in proportion as the facts are more complete.

CHAPTER II.

CARDIOGRAPHY—VENTRICULAR SYSTOLE AND DIASTOLE, THEIR RELATION TO EACH OTHER AND THE HEART'S REVOLUTION—MECHANISM OF THE CARDIAC AND ARTERIAL TRACES—THE NEGATIVE ARTERIAL TRACE—THE PRESPHYGMIC INTERVAL.

Cardiography.

CARDIOGRAPHY comprehends the art and science of the instrumental registry of the movements and phases of the heart's revolution. Or, the term may designate either the art of producing, or the science resulting from, such graphic representations. The instrument which writes the changes is properly termed the cardiograph, and the graphic line produced by it, the cardiogram. Cardiography is less known and practised than sphygmography; nevertheless its history is not unworthy of attention.

The first successful auto-records of the living heart were obtained in 1860 by Chauvean and Marey in their brilliant series of experiments on the horse. These experiments have become classical, and their *technique* need not be detailed here. Suffice it to state that the variations of the blood-pressure in the cavities of the heart and in the aorta, as also the impressions of the heart against the chest-wall, were faithfully and simultaneously recorded during the heart's revolutions.

In the preceding chapter appear eight cardiograms, each embracing six or seven full pulsations taken from a healthy man, aged thirty-two years, with chronograms in fifths of a second written at the same time. The heart, tested by palpation, auscultation, the cardiometer, and cardiograph, was found normal in force, sounds, amplitude, and rhythm; and by comparison with many others it was found to fairly represent in its movements the average of healthy human cardiac pulsations. They are regarded as typical, and they are commended as showing the true systolic and diastolic relationships. Four other cardiac tracings—two each from K and L—also appear. These, though less finely delineated in the smaller waves than those just named, show very distinctly the beginning and end of ventricular systole.

The table on opposite page is a collation of facts derived from measurements on the chronograms of pulsations from the above-named cardiograms. Of the eight, two consecutive pulsations from each were measured. Of the K and L cardiograms, one pulsation from each was measured. In Nos. 1, 2, 3, 5, 6, and 8, the measurements are from contiguous pulsations on either side of line B; in Nos. 4 and 7 they are from consecutive pulsations following the line B; and in Nos. 16 and 23, from single pulsations following the line B on either side of the reversal lines.

The data of the table indicate the truth of the following propositions:

1. In the normal heart the rhythm of its movements is continually changing, and the change pertains to the duration of its entire revolution, and to that of both systole and diastole.

2. The duration of the diastole changes in much greater degree than that of systole.

3. A longer systole may go with a shorter pulsation; and a shorter systole with a longer pulsation.

4. Invariably a longer diastole goes with a longer pulsation, and a shorter diastole with a shorter pulsation.

5. In the normal heart, beating at about 75 to the minute; the average ratio of systole to diastole is very nearly as two to three.

CARDIOGRAPHY.

Table showing, in decimals of a second, measurements of the normal human heart's

		Systole,	Diastole,	Pulsation,	and Frequency per Minute.
No. 1	1st pulsation	.347	.453	.800	75
	2d "	.332	.490	.822	73—
No. 2	1st pulsation	.312	.453	.765	78.4
	2d "	.312	.453	.765	78.4
No. 3	1st pulsation	.336	.443	.779	77+
	2d "	.313	.480	.793	75.6
No. 4	1st pulsation	.328	.518	.846	71—
	2d "	.314	.514	.828	72.4
No. 5	1st pulsation	.322	.487	.809	74+
	2d "	.331	.514	.845	71
No. 6	1st pulsation	.309	.505	.814	73.7
	2d "	.329	.538	.867	69
No. 7	1st pulsation	.344	.454	.798	75+
	2d "	.339	.481	.820	73+
No. 8	1st pulsation	.324	.447	.771	77.8
	2d "	.331	.434	.765	78.4
Average of the 16 pulsations.		.3264	.4786	.805	74.5
K No. 16	1st pulsation	.336	.400	.736	81.5
	2d "	.315	.394	.709	84.6
L No. 23	1st pulsation	.316	.595	.911	65.8
	2d "	.343	.542	.885	67.8
Average of the 4 pulsations.		.3275	.4827	.810	74.9
Average of the 20 pulsations.		.3269	.4806	.8075	74.7

Mechanism of the Cardiac Trace.

In the cardiac cycle there occur as palpable phenomena—

(1) Auricular systole, with contraction of the auricular fibres, hardening of the auricular parietes, sudden rise of intra-auricular blood pressure, impulsion of blood through the auriculo-ventricular orifice, shrinkage of the auricular volume, and systole quickly changing into—

(2) Auricular diastole, with relaxation of the fibres, softening of the parietes, sudden fall to negative of the blood pressure, filling of the cavity from the great veins, gradual increase of the blood pressure, swelling of the volume.

(3) Ventricular systole, with contraction of the fibres, hardening of the parietes, sudden rise of the blood pressure,

closure and tension of the auriculo-ventricular valve, opening of the semilunar valves, impulsion of blood through the arterial orifice, reduction of volume, variable maintenance (after the first change) of contraction, hardness, and pressure, and systole suddenly changing into—

(4) Ventricular diastole, with relaxation of the fibres, softening of the parietes, sudden fall to negative of the blood pressure, opening of the auriculo-ventricular valve, closure of the semilunar valves, filling of the cavity from the auricle, gradual rise by stages of the blood pressure. augmentation by stages of the volume.

(5) The changes and conditions of arterial filling and emptying.

The elements which determine mainly the formation of the cardiac traces are :—

(1) Changes of contraction and relaxation of the cardiac fibres.

(2) Changes of intra-cardiac blood pressure.

(3) Changes of consistence of the cardiac walls.

(4) Changes of volume of the heart.

The cardiac trace is produced in man by the instrumental registry of the variations of protrusion and recession of the intercostal space, occasioned by the action and states of the heart beneath. In general terms the space advances and remains more or less elevated during ventricular systole. while it recedes and remains more or less depressed during ventricular diastole.

We will now endeavor to discern the normal types of the individual factors above stated, whose composition must form the true normal cardiogram.

(1) Systole of the ventricle is marked by contraction of the fibres of the ventricle. At the end of the systole fibres relax, and diastole is marked by relaxation of the ventricular fibres. It is well known that the ventricle continues for a time to alternately contract and relax, even when its cavity is deprived entirely of blood ; yet in normal action it is true that the force and duration of contraction are influenced by the

resistance to be overcome. Experiment shows that a stimulated muscle contracts stronger and longer when weighted. So the ventricle will modify its force of contraction by the resistance to the exit of its blood, and will continue its action, within limits, until the object of its systole, the discharge of an adequate volume of blood into the artery, has been fulfilled. Accordingly, then, it is conceived that contraction begins and increases its force until the semilunar valves open and the blood begins to escape. As the blood passes out into the relaxed aorta the resistance diminishes, and with this the force of contraction diminishes; the aorta becoming filled the resistance increases, and with this the force of contraction increases. It seems probable that the arterial resistance is carried back to the ventricle and stimulates its renewed contraction in the form of a double effort, the last one the stronger. Systole being completed, contraction suddenly changes to complete relaxation of the fibres, and the flaccid condition continues throughout ventricular diastole. The module, then, of systolic contraction of the ventricle would be a line first ascending quickly to a high point, then turning a little downwards, then forming a small upward wave, then a wave running as high at least as the level of the first summit; and the module ,of diastolic relaxation of the ventricle would be a line quickly descending from the high level of the end to the level of the beginning of systolic contraction, and thence running horizontally to the end of diastole. This type is represented in the plate, No. I., by the line beginning with *a b* and continued by the uppermost dotted line to *a'*.

(2) The module of intra-ventricular blood pressure is, in systole, closely that of ventricular contraction. Pressure and contraction must begin simultaneously (although the latter is the initial condition), and the elevation of the one and force of the other proceed *pari passu*, and the climaxes are reached at the same time. After the opening of the valves to the end of systole the blood pressure first falls sharply, then is checked and raised a little by the establish-

ment of arterial tension, and then raised more decidedly by the increased force of contraction engendered by the arterial resistance. Thus in strict comparison the waves of blood pressure are sharper than those of contraction, the first fall and second rise of the former *leads* the weakening and strengthening of the latter, while the renewed contraction *leads* in turn the third and last rise of blood-pressure. In diastole the blood pressure first descends below zero,

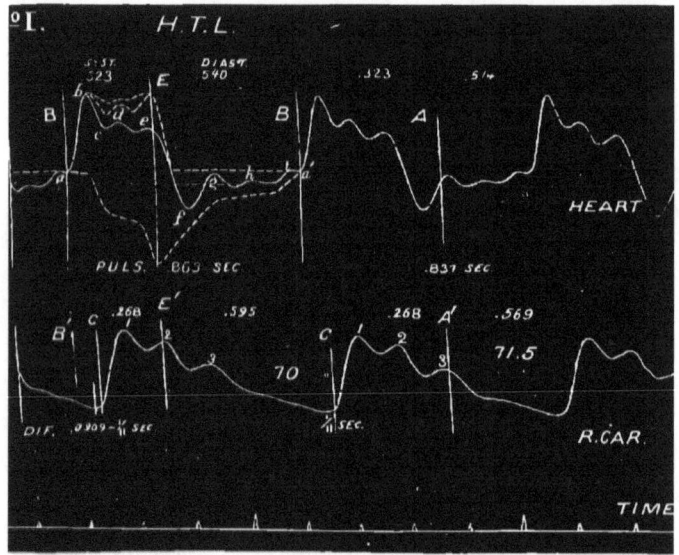

then rises by the influx of blood from the auricle, then more slowly by the continued but slower flow, and then more rapidly again by the auricular charge. The type, then, of the variations of intra-ventricular blood pressure would be, in systole, a line beginning as with *a b*, then running by the second dotted line to the point intersected by *E*, and, in diastole, continued by *e f*, and thence on to *a'*, nearly by the course of the solid middle line there seen.

(3) The changes of consistence of the ventricle are a

varying hardness during systole and a varying softness during diastole. The varying hardness of systole is the result of changes of contraction and changes of blood pressure combined, and its module is *essentially* that of changes of blood-pressure. The varying softness of diastole is the result of relaxation and changes of blood pressure combined, and its module is *purely* that of changes of blood-pressure. So the line of changes of consistence of the ventricle would be essentially the same as that of variations of intra-ventricular blood pressure before described.

(4) The changes of volume of the ventricle are reduction by stages during the sphygmic portion of systole and augmentation by stages during diastole. After the beginning of contraction the ventricle remains at its maximum volume until the blood begins to flow from its cavity. The escape, at first free, permits, at first, rapid diminution; the escape, retarded by the arterial filling, checks the rate of diminution; and the escape again accelerated by the renewed vigor of contraction increases again the rate of reduction. Contraction changing to relaxation, the ventricle suddenly enlarges by lengthening of its fibres and elasticity of its walls, and quick falling into it of blood from the auricle; then the enlargement is slowed, for a time following a slower influx, and next is accelerated again by the charge of auricular systole. Thus the module of changes of volume of the ventricle is typified, in systole, by a line first horizontal, then quickly descending, then more slowly descending, and then again more rapidly descending; in diastole, first quickly, then more slowly, and then again more rapidly ascending. This type is shown in the lowest dotted line running from a to a'.

Such is the analysis and representation of the phenomena of the movements bearing upon the formation of the cardiac trace. Are these modules correct? They are consistent with accepted physiology of the heart's movements; they are supported in principal features, as self-evident propositions, and for the most part they have been confirmed by experiment.* With regard, however, to the intermediate systolic wave, it is admitted that other agencies might be adduced as auxiliary factors in its production, and that these agencies might sometimes cause the wave to present in multiple form. These agencies are: (*a*) locomotion of the heart by back action as the issuing blood meets with resistance; (*b*) locomotion of the heart by lateral pressure of the pulse wave as it distends the descending aorta; and (*c*) vibration of the ventricular parietes under the systolic effort and arterial resistance. These factors, if operative, would usually accord in time with the resistance reflected by the interior, and so contribute to the production of the intermediate wave as ordinarily seen; but if, peradventure, they should be discordant, then the wave might appear multiple, as sometimes seen, instead of single.

Accepting, then, the representations given as substantially true, we proceed to seek in the synthesis of these lines the normal type of the movements of the heart.

We have seen that the modules of contraction, consistence, and pressure resolve themselves essentially into that of intra-cardiac pressure; so the composition sought is substantially between the two lines of pressure and volume, which so far simplifies the investigation.

In the first part of systole—that is, during the pre-sphygmic portion, there being no volumetric decline, the intra-ventricular pressure rises without limitation from this cause to its high point of climax, but from this point the line of systole is lower than the line of pressure would make it, in consequence of reduction of ventricular volume; in the oscillations of blood pressure each decline and each elevation mark a lower position, and the level of the end of systole is much below that of the first summit. The diastolic line of recession is shortened in the upper part, and the rest of the line corresponds with that of pressure. The line, then, we would construct from these elements as the module of the heart's pulsation is the solid line, *a, b, c, d, e, f, g, h, i, a'*, and this is

* See "La Method de Graphique," p. 386.

a real trace of a normal human heart, enlarged to double size by photography.

The module of auricular systole evidently would be composed of contraction of the auricular fibres and intra-auricular blood pressure, and would manifest itself at the surface of the auricle as a single rising and falling wave, appearing a little before the advent of ventricular systole, but over the ventricle this wave would divide itself into two small ones, the first representing the beginning of contraction, and the second the impulsion of blood into the ventricle. Thus the waves h i stand respectively for these events.

We maintain that this division of the effect of auricular systole at the surface over the ventricle is favored in sound theory, and the constancy of these small waves in all well-taken cardiograms is confirmatory proof that such is the mechanism of their production.

This cardiogram, taken from a man in health, aged twenty-eight years, is a trustworthy representation of the movements it followed, and is equal to the best specimens of cardiographic art. It shows correctly the phases and limits of ventricular systole and diastole, and the markings of auricular systole. The valvular closures it does not distinctly show. We know that the auriculo-ventricular valve closes in the first part of systole, and the semilunar valves close in the first part of diastole, but these events are drowned, as it were, in the greater events which include them, and transpire, the one whilst the lever is rapidly ascending, the other whilst it is rapidly descending, thus contravening a distinct record of their occurrence. It sometimes happens, however, that a cardiogram of very sloping descent will show a break which marks the closure of the semilunar valves.

Mechanism of the Arterial Trace.

In the plate the tracing below the cardiac is that of the carotid artery taken simultaneously with the former. It, too, is a typically correct representation of the arterial movements it followed. This trace would differ from a trace of the commencing aorta only in smaller amplitude and a little greater delay on the pulsation of the heart. Allowing only for the latter it may be studied as if it were an actual trace of the initial aorta.

Below is the time in fifths of seconds written by a chronograph also simultaneously with the pulsations. $A\ A'$ are the signal lines made by the levers with the recording slip at rest; B, parallel to A, cuts the basal point of the cardiac trace which marks the beginning of ventricular systole; and C, parallel to A', cuts the basal point of the arterial trace which marks the beginning of the pulse; $B'\ C$ shows the difference in distance between $A\ B$ and $A'\ C$, which is the delay of the pulse on the heart, and measured on the chronogram its time-value is neatly found; also E, parallel to A, cuts the point marking the end of the cardiac systole, and E', parallel to A', cuts the pulse trace where it may, the distance $A'\ E'$ being the same as $A\ E$ of the heart line.

Under the systole of the ventricle at the opening of the aortic valves the blood is launched into and expands the artery. Under the charge the climax of arterial tension is soon reached, but is rapidly receded from in consequence of the quick, outward flow of the blood. Thus is formed the first wave of the pulse and first line of recession. The climax of arterial tension is reflected through the open valve back to the heart where it raises a small wave of pressure and excites the ventricle to renewed contraction. This last effort of the ventricle quickly sends forth more or all of the remaining contents staying the arterial recession, and raising a second wave. Thus is formed the second wave of the pulse. Systole being ended and diastole begun, the pressure in the ventricle becomes negative, which starts the arterial current towards the heart, causing rapid recession of the arterial line. This centripetal flow with arterial fall is duly checked by the closure of the aortic valves, and the rebound of the blood completes a third wave. Thus is formed the second line of recession and third or aortic wave of the pulse.

(This wave is in fact a double wave, the division arising from the shock of closure and rebound of the blood.) After this the pulse gradually recedes by the arterial emptying towards the capillaries, and the descending line is thrown into a series of indistinct wavelets, which, if legibly traced, would probably give one for the ventricular filling, one each for auricular contraction and impulsion, and one for the first phase of ventricular contraction, these influences of course transmitted to the closed aorta.

Correspondences between the Phases of Cardiac and Arterial Movements.

The arterial line is yet declining when the cardiac ascent begins. This is demonstrated in the tracings in that B, which marks the basal point of the heart trace, cuts the pulse trace at a considerable distance anterior to C, the basal point of the latter. It is known that the time occupied by the transmission of the pulse-wave from the aortic orifice to the carotid point is only a small part of the time represented by the distance B' C. (The division is indicated by the dotted line between B' and C.)

The high point of the cardiac trace corresponds closely with the basal point of the arterial. This also is demonstrated in the tracings. The distance Bb is found to equal the distance B' C minus the small distance representing the transit time. Next the arterial line ascends while the cardiac descends, and there is more or less correspondence between the arterial summit and cardiac depression c. The first arterial depression and second wave of the pulse correspond with the remaining part, cdc, of cardiac systole. The intermediate cardiac wave d fails to appear in the pulse trace for the reason, probably, that as it is merely a small reflected wave in the heart, its second reflection in the pulse is not apparent. The renewed contraction of the heart would raise a corresponding arterial wave, the second wave of the pulse. This correspondence is shown by the relation of the lines

EE' to these phases, respectively. After this correspondences are no longer distinctly traced; for although the second arterial notch plainly marks the closure of the aortic valves, this event, as before stated, is not shown in the rapidly falling diastolic line of the heart, and, as previously remarked, the waves of ventricular filling, auricular contraction, auricular impulsion, and initial phase of ventricular contraction, all indicated in the cardiac trace, have but faint and uncertain correspondences in the arterial.

Taking the rise of the pulse as its beginning, the cardiac systolic phase is shorter than the cardiac systole, and the cardiac diastolic phase is longer than the cardiac diastole, in either case by so much as the pertaining presphygmic interval.

It is proper to state that the correspondence between the end of ventricular systole and the acme of the second wave of the pulse, although complete in this showing, is liable in other instances to be less exact; the default, however, is never sufficient in frequency or extent to raise a doubt as to the correlation between these events. It sometimes happens that the second pulse-wave attains its point before the end of systole is accomplished, as if the ventricle expelled its blood and then held its contraction a moment longer. On the other hand, it would seem that the ventricle sometimes changes from systole to diastole before its emptying is completed, and the second wave is sent forward a little delayed on the systolic ending.

Other Demonstrations.

The illustration before us is a true map, in which are correctly shown all the facts of the cardiac and arterial cycles here inscribed, their characters, phases, and relations to each other, and in which the chronometry of the events recorded are shown with mathematical precision.

In addition to phenomena already considered, we instance the following showings of durations:

CARDIOGRAPHY. 67

First cardiac cycle, .863 second ; systole, .323 ; diastole, .540.
Second cardiac cycle, .837 second ; systole, .323 ; diastole, .514.
First cardio-carotid interval, .0909 = 1-11 second.
Second cardio-carotid interval, .0909 = 1-11 second.
First ventricular presphygmic interval, .0555 = 1-18 second.
Second ventricular presphygmic interval, .0555 = 1-18 second.
First cardiac systolic phase of pulse, .268 second ; cardiac diastolic phase, .595 second.
Second cardiac systolic phase of pulse, .268 second ; cardiac diastolic phase, .569 second.

The duration of the presphygmic interval was approximately determined in the following way: The carotid-femoral interval in the same subject had previously been ascertained to be .0909 second (unusually long), and on the basis of this the transit interval between the heart and carotid was calculated at .0354 second, and the latter deducted from the cardio-carotid time, difference .0909 second, gave .0555 as the presphygmic interval. The near accuracy of the result so obtained will be apparent when it is considered how closely alike, except in lengths, are the arterial tracts concerned, and that such similarity would insure uniformity in the rate of propagation of the pulse-wave, and then with correct measurement of the arterial distances and correct observation of the carotid-femoral interval, there could be no room for material error in the estimation of the cardio-carotid transit interval.

Another point is the precedence of the auricular to ventricular systole. Measuring from the beginning of the first wave, the interval is about two tenths of a second, and from the beginning of the last wave, about one tenth of a second. The last wave is that which is usually assigned as answering to auricular systole; but plainly enough this is too close to the systole of the ventricle to include the beginning of auricular contraction, while the first wave is in proper position to mark this event, and the second wave is properly located for marking the event of auricular impulsion.

In accordance with the foregoing views and demonstrations, as well as proven by numerous careful experiments not here recorded, the cardiac and arterial movements and cardio-arterial time-differences are expressed in the graphic lines of the pulsations as follows:

Of the heart, the initial force of systole (ventricular) is indicated by the height of the main ascent. The quickness of systole by the steepness of the main ascent. The sustentation of systole by the relative height, as respects the first summit, of the point of systolic ending. The duration of systole by the horizontal distance between successive points of systolic beginning and ending. The duration of ventricular diastole, obviously, by the horizontal distance between successive points of systolic ending and beginning. And the duration of the cardiac cycle by the horizontal distance, preferably, between successive systolic beginnings. Of all these phases and qualities of the heart's pulsation, except the initial force of systole, the graphic lines are true indices without qualification. Regarding the exception, the height of the main ascent is much influenced by other conditions besides the force of systole, as the relation of the heart to the chest-wall, the width and thickness of covering of the intercostal spaces, the position of the subject, all of which in different cases so modify the amplitude of the traces that the height of the upstroke can only be received as an index of systolic energy when other things are known to be equal. Thus the index may be of value in the same individual, but for purposes of comparison between different individuals it must be held of little significance.

The auricles, so far as admissible in view of their position, express their energy of systole by the height of the auricular waves, and their time of systole by the distance between the auricular and ventricular rises.

Of the pulse, the amplitude is shown by the height of the major ascent. The quickness by the steepness of the major ascent. The tension by the relative height and horizontal distance of the second wave and aortic notch from the basal point. The dicrotism by the height of the aortic wave. The duration or frequency by the distance between successive basal points. The duration of the cardiac systolic and cardiac diastolic portions by the horizontal distances measuring these divisions. All these qualities of the pulse are thus accurately shown, only in estimating the amplitude by the height of the main ascent regard must be had for the thickness of the tissues over the artery; for plainly of two arteries of the same calibre and movement the more superficial one will afford the higher trace.

Of the cardio-arterial time-differences, these are shown by the difference in distance, from synchronous signals, of the cardiac and arterial basal points, respectively (the synchronous signals are the lines crossing the traces, made by the levers with the recording slide at rest as before described). Thus the cardio-carotid time-differences may be found to measure one twelfth of a second, the cardio-radial one sixth of a second, and the cardio-dorsalis pedis one fourth of a second; or, whatever the interval at the moment of observation between the pulsation of the heart and that of any artery, this differential distance mathematically expresses it.

In further illustration of this subject, and to show variations of form and chronometry of the cardiac and arterial movements in health, tracings from three other individuals are produced. They are in *fac-simile* and with the original or enlarged representation just considered were all taken with the compound sphygmograph.

Nos. II. and III. are respectively from young men each twenty years of age. No. IV. is from a man aged fifty-four years; the first part at rest, the last part immediately after active exercise.

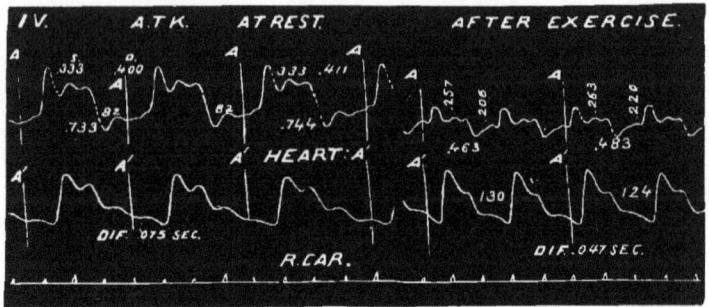

In Fig. 10 the top row of figures shows in decimals of a second the duration and variation of individual cycles, as carefully measured on the chronogram in fifths of seconds, below. Thus we have come to learn that the heart, however regular its action may appear, is subject to an incessant oscillation of rhythm.

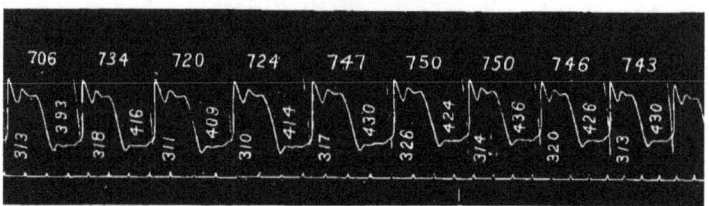

FIG. 10.

Not to dwell on the causes of these rhythmic variations, the fact of a definite relationship between the respiration and variations of the cardiac cycle may here be mentioned. It has been demonstrated that the cycles of the heart are shorter in inspiration and longer in expiration; the shortening running over into the first part of expiration and the lengthening running over into the first part of inspiration. See Fig. 11.

FIG. 11.—RESPIRATION AND HEART TRACED SIMULTANEOUSLY, WITH THE TIME; AND MARKED TO SHOW THE PULSE-RATE OF THE INDIVIDUAL CYCLES.

The Negative Arterial Trace.

First in 1874, when tracing with Marey's sphygmograph the radial pulse of a boy affected with disease of the heart, there was obtained a distinct inverted trace, which was then interpreted as of pathological significance. Subsequently, in practising with his own instrument, the author has constantly encountered this phenomenon, and has come to learn that it is the negative arterial pulse, normally existing, and revealed in an inverted trace when the instrument is applied to the artery in a particular way.

The negative pulse may be demonstrated in all the larger superficial arteries by placing the explorer closely by the side of the artery instead of directly upon it, as when the ordinary positive trace is obtained. The negative trace, compared with the positive, is inverted, and of smaller amplitude, and presents more distinctly the smaller secondary waves (those occurring during the time of closure of the sigmoid valves); also its respiratory undulations are more marked, and usually the trace is best taken at a lower pressure. It corresponds with the positive in general form, in the number and relative position of the secondary waves, in the aortic notch, and is exactly synchronous with it in primary start and each individual phase throughout. In demonstration of these statements the accompanying figures of traces are given.

Fig. 12 shows the carotid, negative and positive. Fig. 13

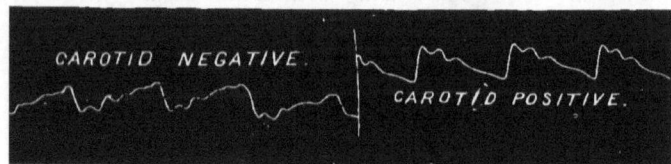

FIG. 12.

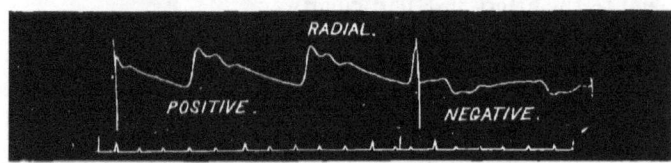

FIG. 13.

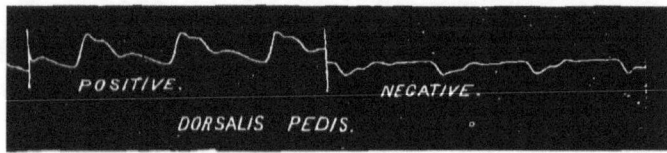

FIG. 14.

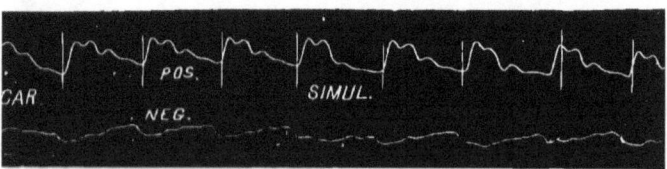

FIG. 15.

shows the radial, positive and negative. Fig. 14 shows the dorsalis pedis, positive and negative. And Fig. 15 shows the same carotid, positive and negative, taken simultaneously.

These traces speak for themselves; with regard, however, to Fig. 15, a word of explanation may be added. To obtain the two kinds of traces from the same carotid simultaneously required a particular adjustment of the explorers, which was found only after repeated trials. The vertical lines cutting the traces were made by the writing levers respectively, with the slide at rest; and, cutting the traces as they do, they demonstrate the exact synchronism between the negative and positive pulses.

Mechanism of the Negative Arterial Pulse.

The mode of production of the negative arterial trace has been a puzzling problem, but it is believed now its solution is found in the locomotion of the arteries. With each cardiac systole, the arteries, in addition to expansile movements, rise *en masse* from their beds, and with each cardiac diastole, they, in addition to retractile movements, fall back to their beds. This change of place must occasion a pushing out of the tissues over the spot towards which the artery moves, and a simultaneous drawing in of the tissues over the spot from which it moves. The locomotion will be in the direction of least resistance, and in the case of superficial arteries is likely to be to one side as well as forwards. If a visible superficial pulse, notably the radial, be watched, the artery appears to roll to one side in systole and fall back in diastole; the contiguous integuments simultaneously rising at one spot and falling at another.

Now, if the explorer be placed fairly over the artery, it receives the force of the locomotion as well as the expansion, and so is pushed up in systole, causing to be registered the positive arterial pulse; whereas, if the explorer be placed by the side of the artery over the spot from which the latter moves in systole, it is affected by an aspiration, and so is drawn down in systole, causing to be registered the negative arterial pulse. Moreover, impingement of the instrument in this situation tends to facilitate the arterial displacement.

The positive trace falls in diastole in consequence of emptying and retreat of the artery, while the negative rises in diastole in consequence of return of the artery to its apposition with the explorer. The secondary waves of the negative traces are the same as in the positive, because the delicate contact of the explorer is sufficient to receive their impression as transmitted along the artery; and the small, diastolic waves are more distinct in the negative because the lighter pressure is more favorable for their reflection, the pressure in the case of the positive trace almost, if not quite, obliterating the arterial lumen in diastole. Also, for the same reason (lighter compression of the artery), the respiratory undulations are more marked in the negative line.

The characteristics given unmistakably stamp the trace under consideration as of arterial origin. The negative venous trace, where obtainable, differs essentially from this in form and in chronometry; besides the vein would be effectively collapsed by the pressure employed to develop this trace.

The demonstration of the negative arterial pulse adds an interesting fact to our knowledge of the physiology of the pulsations, and one whose utility is already apparent, in that its recognition will relieve the study of the negative venous pulse from a source of confusion in fact seriously encountered.

The Negative Cardiac Trace.

This is a proper place to allude to the negative trace of the pulsations of the heart, a phenomenon already recognized, but whose explanation has not been entirely satisfactory.

The negative cardiac trace is usually of less amplitude than the positive, although in some cases it equals or even exceeds it, and in some again only the negative can be obtained. Its main descent begins synchronously with the main ascent of the positive, and it marks the same duration of systole and diastole; the secondary waves also correspond in number and position with those of the positive.

Fig. 16 shows simultaneous traces of the same heart, taken both positively and negatively. The vertical lines show the synchronism between the traces.

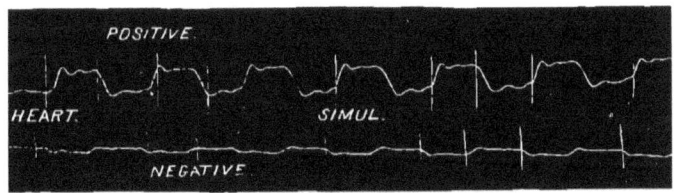

Fig. 16.

With regard to the mechanism of production, it is believed it is similar to that of the negative arterial—namely, by a rhythmical locomotion of the organ. With a certain position of the explorer the heart glides or rolls from under it in systole and returns to it in diastole, causing the negative trace; whereas, when the explorer is placed fairly over the spot of greatest positive impulse the heart presses it in systole and remits the pressure in diastole, causing the positive trace. Cardiac aspiration from reduction of volume, although conceivable as a possible factor, does not of itself afford a sufficient explanation, for cardiac reduction of volume can only begin a notable time after the beginning of systole, whereas the negative cardiac trace begins, as demonstrated, synchronously with systole.

Dr. S. E. Post, after conversion from the theory of venous to that of arterial origin of this trace, has divined and pointed out the agency in its production of "direct impact against the frame instead of against the diaphragm of the explorer."* There can be no question that such "frame-impact" is capable of causing the diaphragm to descend in systole and ascend in diastole, and of reflecting in inverted relation all the details of the arterial pulse.

*Archives of Medicine, February, 1884.

The Presphygmic Interval.

The experimental proof that a cardiogram taken from the chest-wall represents the phases of the cardiac revolution, and that the basal point of the main ascent marks the beginning of cardiac systole, has been furnished by Marey in one of his celebrated experiments upon the horse. Simultaneous inscriptions of the blood pressure in the interior of the right ventricle, and of the heart's action against the chest-wall, showed correspondence with each other in the essential points, and exact synchronism in the beginning of the two up-strokes. Marey also demonstrated, in an experiment of his series on the horse, that the two ventricles begin to contract precisely at the same time ; and in another he states that he has never observed default of synchronism between the two ventricles. Thus we are prepared to accept the cardiograms obtained from the chest-wall of man as correct representations of the movements of the human heart, and accept the basal point of the main ascending line as marking the beginning of ventricular systole. In yet another experiment of Marey's series on the horse, one sound was lodged in the left ventricle, and another in the aorta just above the valves. The inscriptions thus obtained showed that the blood pressure in the ventricle began to rise a notable time before the beginning of the rise in the aorta. The interval between these beginnings represents the time elapsing after the ventricle begins to contract until the aortic valves open and the blood begins to escape into the artery. This is the presphygmic portion of the systole of the ventricle, or the *Presphygmic interval*, as it may be termed,—the *syspasis* of A. H. Garrod.*

As it has been proved that the pulse-wave rises later in an artery removed from the heart, evidently observations between the heart and an arterial point beyond the root of the aorta will give a time difference made up of the presphygmic interval and the time required for the pulse-wave to travel

---

* Proc. Royal Society, No. 157, 1875, p. 144.

from the aortic orifice to the point designated, the transit time of the pulse wave.

To determine the duration of the presphygmic interval in man, simultaneous tracings of the heart and carotid pulse are obtained with an accompanying chronogram, or in the presence of a known velocity of the surface receiving the inscriptions, and the time difference thus ascertained is noted. Next, the transit time between the aortic orifice and carotid point is determined. The latter time deducted from the former gives the presphygmic interval.

The duration of the heart-carotid interval in man has been investigated. Czermak,* in 1864, by means of his photosphygmograph, measured the interval at .087 second. Mosso,† by means of a cardiograph for the heart and a tambour devise for the carotid pulse, measured the interval at between ten and eleven hundredths of a second.

The author has made a very large number of observations and measurements of the heart-carotid interval with the following results:

1. That the interval is subject to considerable variation in the same individual, and in different individuals compared with each other.—See Figs. 18, 19, 20.

2. That the interval varies inversely with the pulse rate, directly with the pulse duration.—See Figs. 18, 20, 21—all from the same person.

3. That the average duration of the interval with pulse at 75 to the minute, four fifths of a second long, is about .08 second.

According to these data, a pulse of 60, one second long, will give the interval one tenth of a second, and the rule may be formulated that the average cardio-carotid interval is one tenth the duration of the pertaining pulsation. This rule is approximate only, for the interval varies from other factors besides pulse frequency; nevertheless, it will be found of practical value in determining approximately what should be a standard time difference for a given pulse rate.

* *Med. Chir. Rev.*, January-April, 1865.
† "Die Diagnostik des Pulses," Leipzig, 1879, p. 45.

Accepting, then, .08 as a standard heart-carotid interval in health, with a pulse at 75 per minute in pursuance of our investigation, we now seek a standard transit time of the pulse-wave over the arterial tract included. As the aortic root is not accessible for observation in man, obviously the desired data must be arrived at indirectly. This is the method: The velocity of the pulse-waves over other arterial tracts is ascertained, and if these are found to differ, the velocity over the arterial tract most nearly corresponding with that between the aortic root and carotid point is selected as the basis for computing the cardio-carotid transit time.

If any trustworthy observations could be found on the velocity of the pulse-wave along different portions of the arterial tree, they would be presented here. In default of such findings, recourse has been had to personal experiments which have demonstrated the following facts:

The mean average velocity of the pulse-wave from the carotid point to the dorsalis pedis is 361 inches per second; from the carotid to the radial 290 inches per second; from the carotid to the femoral 269 inches per second; from the femoral to the dorsalis pedis 464 inches per second.

Thus the pulse-wave velocity over different arterial tracts is shown to vary in an important sense. The tract between the carotid and femoral is most closely allied to the tract between the aortic root and carotid; the similarity, indeed, is very intimate; accordingly, we utilize for our purpose the carotid-femoral pulse-wave velocity.

The measurement of the arterial length between the aortic root and carotid point is placed at 7 inches. Hence $\frac{7}{269} = .026$ second expresses the duration of the interval sought.

The pulse-wave velocity, as has been determined, though differing considerably in different individuals, in the same individual is subject only to a limited variation, even under considerable changes of the circulation, and variations of pulse frequency exert upon it no appreciable disturbance.

(The latter fact has also been demonstrated by Garrod.) Therefore, for the short distance of 7 inches, .026 second may stand as an approximate quantity to express the cardio-carotid transit interval for any pulse rate or order of conditions, aortic aneurism alone excepted.

These data give the formula: Cardio-carotid interval, with pulse at 75, .08 second, less the cardio-carotid transit interval .026 second, equals the presphygmic interval .054 second. Then, in any case, physiological or clinical, aneurism within the included tract excepted, to find the presphygmic interval it is only required to find the heart-carotid interval and deduct from it .026 second.

But, inasmuch as the presphygmic varies largely, and the transit for the short arterial length in question inappreciably (aneurism excepted), the cardio-carotid interval practically represents the presphygmic.

Garrod, studying this subject from a basis of experimental observation of the cardiac and radial pulsations, arrived at the theoretical conclusions that the cardiac systole varies inversely as the square root of the pulse rate, and the cardiac systolic portion of the pulse varies inversely as the cube root of the pulse rate; the operations of these rules being to reduce rapidly the presphygmic interval as the pulse increases in frequency, and at 170 to render it *nil*. Garrod's results from his equations make the presphygmic intervals much shorter throughout, and the rate of reduction more rapid than the author's.

Observations do not, as yet, permit the determination of the limit of pulse frequency on the one hand, and of infrequency on the other, at which the cardio-carotid interval ceases to vary, as enunciated, inversely with the pulse rate: between 120 and 60 it is quite sure the rule holds; beyond these there appears to be uncertainty.

By the rule of reduction the presphygmic interval would become *nil* at 230, and at 180 it would be $\frac{1}{1000}$ of a second; yet it is scarcely possible that it should become *nil* even if the former rate could be reached, and the probability is that the shortest permissible interval would exceed that just mentioned.

The variations of the presphygmic interval will bear further elaboration. It often varies in the same healthy person at the same sitting with the same pulse rate; but the range under such circumstances is limited. The explanation of this variation is found in the incessant minor changes in the manner of the heart's movements and relative changes of blood pressure in the ventricle and aorta: at one systole the contraction is quicker, shortening the interval; at another, it is slower, lengthening the interval; at one ventricular discharge the arterial pressure is higher, delaying the opening of the valves; at another it is lower, precipitating the opening of the valves. The reality of these incessant changes is exemplified in the cardiac rhythmic changes taking place,—changes in successive pulsations and successive systoles and diastoles in hearts whose action appeared the most regular.

The interval varies more notably in different healthy individuals with the same pulse rate. This fact finds its explanation in individual characteristics, the same as a pulse rate, or pulse form, or quality of the heart's action, peculiar to different persons.

But as before presented the great variations are co-relative with the variations of pulse rate. A difference of a few beats per minute may not cause a difference of interval sufficient for certain measurement by present means; yet, other things being equal, it is probable that any change of pulse frequency changes the presphygmic interval. A variation of ten per minute will almost always show a notable difference in the interval. (See Fig. 20.)

This correspondence between the presphygmic interval and the rate of the pulse is an interesting and significant fact, and one of a distinctive group we had never known but for the precision of the graphic method. In explanation, the interval shortens as the pulsations shorten, because: (1) The abbreviation of the cardiac pulsation tends to abbreviate

FIG. 18.

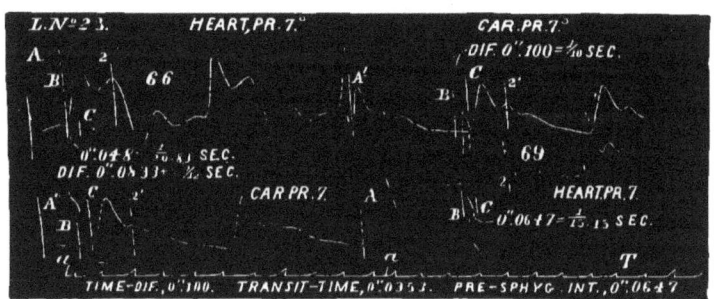

FIG. 19.

all its component parts. The length of diastole is markedly diminished as the pulse rate is increased, and the length of systole, although not so regularly corresponding as diastole to moderate variations of pulse rhythm, nevertheless, is found shorter in frequent than in infrequent pulses. The presphygmic interval, being a component part of the cardiac pulsation, is naturally proportional to its duration. (2) In frequent pulsations the cardiac muscle contracts quicker, and more rapidly raises the ventricular blood pressure to the point of forcing the aortic valves. (3) The accelerated blood current in its passage through the heart, with the usually free passage through the capillaries, constantly tends to surcharge the ventricle and deplete the arteries, thus favoring the earlier opening of the valves.

Explanation of Plates.

The space between the lines *B C* shows the difference of distance between the basal points of the respective pulsations and the signal lines *A A;* the figures near express the value of this difference in fractions of a second as measured on the chronogram below. The figures placed above a pulsation indicate its rate per minute. In Figs. 20 and 21 time differences are expressed by figures placed below the distal basal point without indicatory lines. In Figs. 18 and 19 the tracings are by reversal of the bases to prove the accuracy of the instrument. Other features are self-explanatory.

CHAPTER III.

PULSE-WAVE VELOCITY.

An Experimental Inquiry into the Causes of the Variations of Pulse-Wave Velocity.

"In every truth there is utility, either at hand or among the certainties of the future."—PAGET.

OUR knowledge of velocity of the pulse-wave has all been acquired within a comparatively recent period, and still later are our acquisitions in regard to the interval between the beginning of ventricular systole and rise of the aortic pulse. Previous to the present investigation, it had been determined that the pulse-wave velocity and duration of the cardio-aortic interval were both subject to variations, but the real causes of the variations remained to be determined by experimental inquiry and observation.

The experiments were made with the author's apparatus for simultaneous inscriptions, which, as known, differs from Marey's, notably in transmitting by water instead of air. This mechanism, on account of convenience and accuracy, was found admirably adapted to the work in hand. Also the author's accustomed method was followed in the measurements of the time intervals, and preparation of the slides for illustration.

Whilst the movements were being written, the chronograph also wrote the time in fifths of seconds. Immediately after an experiment, the slide was repassed, and while halted at selected points, the levers, by manipulating the membrane or tube of the explorers, were made to describe their curves across the line of traces; and it was usually arranged so that the proximal lever would cross at or near the beginning of the waves, in order to simplify the measurements and render the time-relations more apparent to the eye. These lines are synchronous signals cutting the traces at the same instant, and always indicate the exact time-relation to each other of the movements recorded.

The measurements were carefully made on the slides, by means of a transparent isinglass scale, ruled in 1-100ths of an inch. With this measure and the chronogram, and a magnifying glass, it was easy to compute the time in fine fractions of a second. This was done and the results written on the slides.

The engravings are faithful reproductions throughout. The lines were photographed on wood, using the glass slides as negatives, and then the engraver, with great care and skill, followed the lines without deviation. So perfect is this work, that the fine measurements made on the glasses are found to hold good in the reproductions.

The schema employed was very simple, yet satisfactory in its workings. (See diagram.) The tube or system conducting the liquid waves connected at either end with one of the horizontal limbs of a hollow T piece. The opposite limb of the proximal T connected with a tube twelve inches long, leading to the egress neck of an elastic bulb or pump. The ingress neck of the pump connected with a tube of convenient length leading to a reservoir. The opposite limb of the distal T connected with a flexible tube, three feet long, provided with a stop-cock one foot from its outer end, and communicating near its middle with a water manometer graduated in inches. The vertical limbs of the T's communicated with the transmission tubes of the recording apparatus, through the media of circular chambers, each divided transversely by a delicate elastic diaphragm; said receptacles being substituted for the ordinary explorers, and here termed "receivers."

Thus the reservoirs supplied the liquid; the pump sent

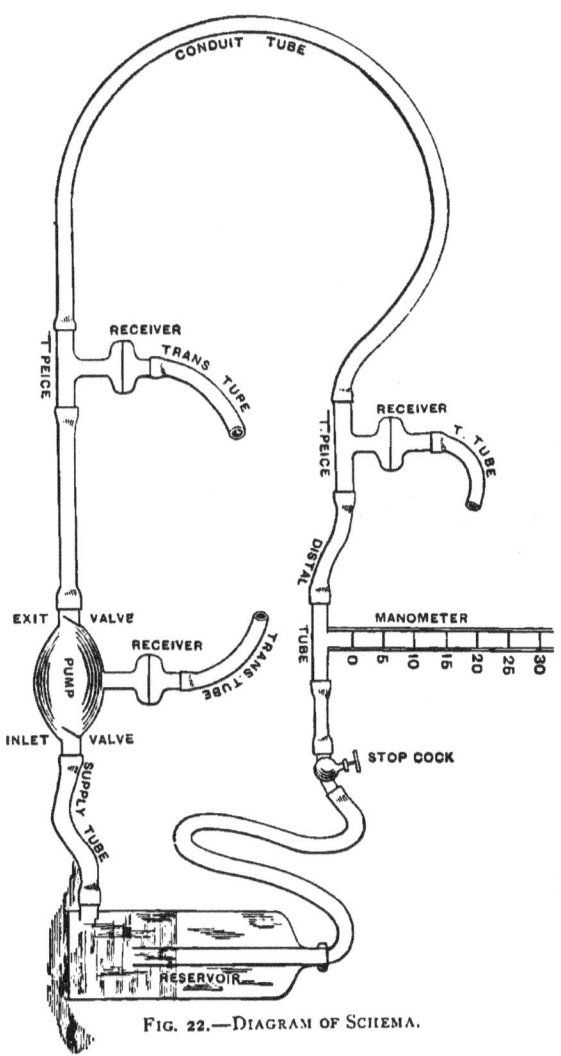

Fig. 22.—Diagram of Schema.

waves along the conduits; the stop-cock regulated the discharge; the manometer measured the pressure; the T pieces and receivers permitted the waves to act directly upon the recording mechanism, and the latter registered their form and time as they manifested at the points of observation. The exit tube could be placed to discharge again into the reservoir, or a separate vessel. The pressure could be raised to any height by working the pump with the exit stop-cock closed, or partially so. A current as swift as desired could be created through the tubes by elevating the reservoir and opening the exit tube placed to discharge at a lower level. Different conduit tubes could readily be placed between the T pieces. The form of the waves could easily be varied by the manner of working the pump. The conditions could be changed at will, and the graphic apparatus would faithfully record the results.

It will be observed that the experiments on liquid waves in elastic tubes differ from Marcy's and others—first, in the character of the graphic apparatus; second, in transmitting the waves directly from the interior of the schema; and third, in the close showings and estimates of the time intervals and wave velocities as manifesting under a variety of conditions.

These experiments were essential to the success and completeness of results aimed at in the investigation; and, as will be seen, the results, when comparable with those of other experimenters, are sometimes confirmatory, other times contradictory.

Problem 1.—To determine the influence of tubes of different degrees of stiffness or elasticity on the velocity of the liquid waves sent along their interior.

First Experiment.—A glass tube, three-sixteenth inch bore, and six feet long, bent in ⊂ form, was placed in communication with the two T's of the schema. The water in the reservoir was at such level that the pressure in the tube measured four inches by the manometer. The exit stop-cock was open, and the pump worked rhythmically by the hand with medium quickness and force. The graphic apparatus in order, with smoked glass in position and chronograph running, the carriage was started, and traces of the waves simultaneously obtained from the two points. Immediately the carriage holding the slide was passed through again, halting it where the upper level would be opposite, or nearly, the beginning points of the proximal waves, and then striking both levers across the line of the traces. These lines made by the levers serve to indicate the distance between the beginnings of the proximal and distal waves, and this distance measured* on the corresponding part of the chronogram gives the time difference between them. In this experiment, Fig. 23 shows the result. The retardation of the distal waves on the proximal is seen to be extremely short. It seems to measure about $1\frac{1}{2}$ hundredths

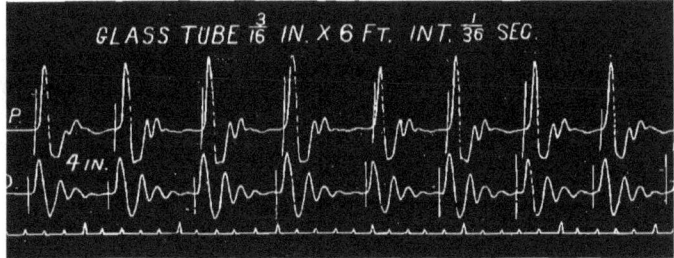

FIG. 23.—GLASS TUBE, $\frac{3}{16}$ INCH BORE, 6 FEET LONG; INTERVAL MEASURED AT $\frac{1}{36}$ SECOND.

of an inch, while the fifths of seconds measure 11 hundredths, which would give a time difference of $\frac{1}{360}$ second, and a wave velocity over the six feet of distance of 216 feet per second.

Second Experiment.—A thick, firm rubber tube of the same bore and length was substituted for the glass, and the experiment under all other conditions conducted precisely as in the preceding. The result is shown in Fig. 24. The measurement is placed at $\frac{2}{11}$ of $\frac{1}{5} = \frac{1}{27.5}$ second, and which gives a wave velocity of 165 feet per second.

* Measurements are readily made by hair dividers and a hundredths inch scale, but more conveniently still by the use of a transparent scale of fine divisions.

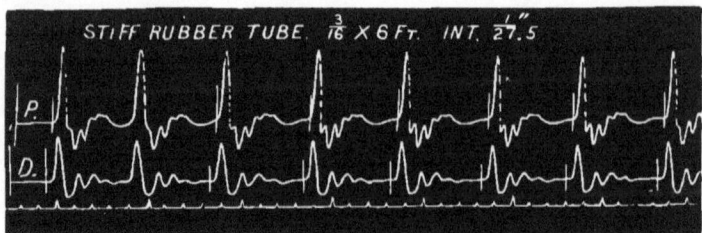

FIG. 24.—STIFF RUBBER TUBE, $\frac{3}{16}$ INCH BORE, 6 FEET LONG; INTERVAL $\frac{1}{27.5}$ SECOND.

Third Experiment.—A softer and more yielding tube of the same bore and length was employed. Fig. 25 shows the result. Formula, $\frac{5}{12}$ of $\frac{1}{5} = \frac{1}{12}$ second; wave velocity 72 feet per second.

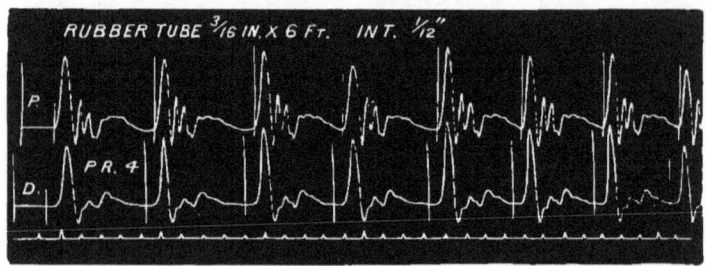

FIG. 25.—ORDINARY RUBBER TUBE, $\frac{3}{16}$ INCH BORE, 6 FEET LONG; INTERVAL $\frac{1}{12}$ SECOND.

Fourth Experiment.—A lighter tube than last, $\frac{1}{8}$ inch bore and same length, employed. Result in Fig. 26. Formula $\frac{7}{12}$ of $\frac{1}{5} = \frac{1}{8.5}$ second; wave velocity 51 feet per second.

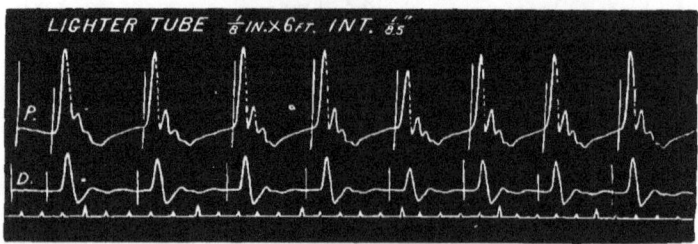

FIG. 26.—LIGHTER TUBE, $\frac{1}{8}$ INCH BORE, 6 FEET LONG; INTERVAL $\frac{1}{8.5}$ SECOND.

Fifth Experiment.—A tube of last description was made very soft and lax, and expanded to $\frac{3}{16}$ inch bore, by steeping in gasoline, and then put in experiment. The first pair of waves of Fig. 27, traced under parallel conditions with the others, shows the result. Formula, $\frac{11}{23}$ of $\frac{2}{5} = \frac{1}{5.2}$ second; wave velocity 31 feet per second.

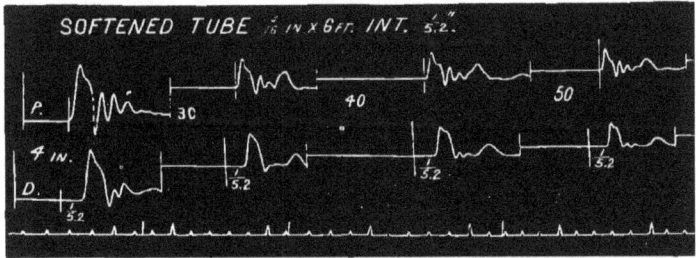

FIG. 27.—TUBE SOFTENED IN GASOLINE, $\frac{3}{16}$ INCH BORE, 6 FEET LONG; INTERVAL $\frac{1}{5.2}$ SECOND.

Sixth Experiment.—Not being able to find in the market tubes of the thinness desired, one was prepared from a strip of delicate rubber cloth, by cementing the edges. This tube, $\frac{3}{16}$ inch bore and 2 feet long, placed in experiment, gave Fig. 28 as result. Formula, time difference $\frac{1}{14}$ second for 2 feet, would make $\frac{3}{14}$ second for 6 feet; wave velocity 28 feet per second.

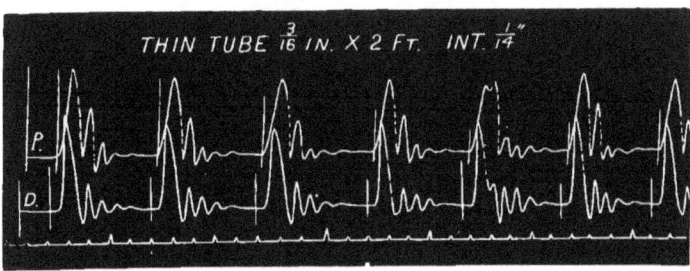

FIG. 28.—THIN RUBBER TUBE, $\frac{3}{16}$ INCH BORE, 2 FEET LONG; INTERVAL $\frac{1}{14}$ SECOND FOR 2 FEET.

Seventh Experiment.— A chicken's intestine, averaging ¼ inch diameter and 2 feet long, placed in experiment, gave Fig. 29. Time difference ¼ second for 2 feet equal to ¾ second for 6 feet; wave velocity 16 feet per second.

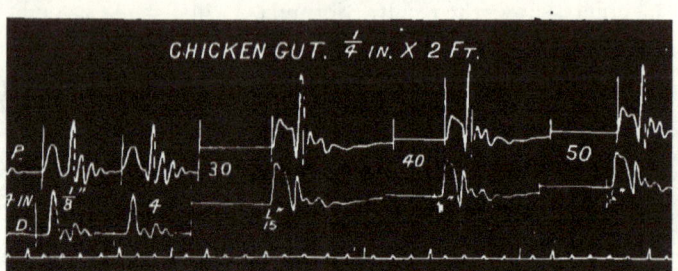

Fig. 29.—Chicken's Intestines, Average ¼ Inch Diameter, 2 Feet Long. Int. ¼ Sec. = ¾ Sec. for 6 Feet.

Eighth experiment.—A calf's aorta, averaging ½ inch diameter, and 18 inches long, with the branches tied, gave Fig. 30. Time difference $\frac{1}{8.5}$, or $\frac{2}{17}$ second for 1½ feet, equal to $\frac{8}{17}$ second for 6 feet; wave velocity 12.75 feet per second. This arterial tube was very soft and elastic, but apparently firmer than the chicken's intestine. We shall see further on that size of tube is an important factor of modification.

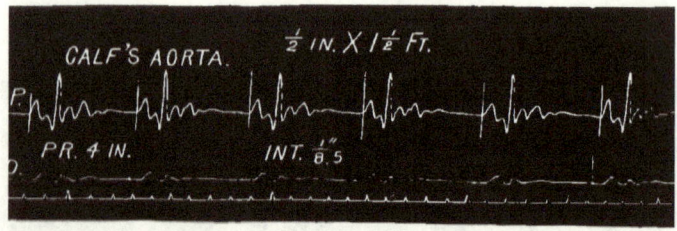

Fig. 30.—Calf's Aorta, Average ½ Inch Diameter, 1½ Feet Long; Int. $\frac{1}{8.5}$ Sec., $\frac{4}{8.5}$ Sec. for 6 Feet.

These experiments demonstrate in a specific manner, that the velocity of liquid waves in elastic tubes is proportional directly to the stiffness, inversely to the elasticity of the tube traversed. And as bearing upon the rate of pulse propagation in living arteries, they indicate the important modifying influence which the state of the arterial walls as to stiffness or elasticity must exert upon the same. In the next chapter will be given actual verifications from man of a positive ratio between the velocity of the pulse-wave and degree of arterial stiffness.

Problem 2.—To determine whether the velocity of liquid waves in the interior of elastic tubes is modified by the mode of impulsion.

Experiments.—Under stated conditions whilst tracings were being taken, the pump for one part of the run was worked with quick, and for another part with slow, impulsion. Different tubes were employed and tracings taken at various pressures, and many experiments made, but always with the same negative result. Fig. 31 was taken with the chicken gut, before described, at 10 inches pressure. The time difference measures the same, viz.: $\frac{1}{10}$ second, under the opposite modes of impulsion. Another illustration of this fact is shown in Fig. 33.

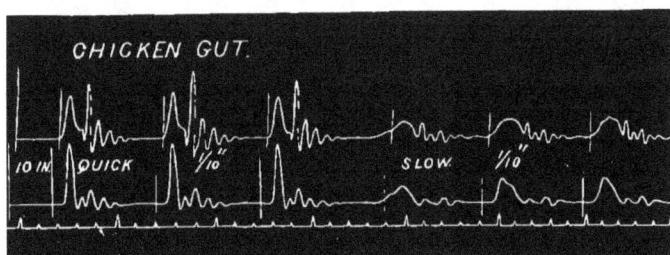

FIG. 31.—CHICKEN GUT, SHOWING NEGATIVE EFFECT OF OPPOSITE MODES OF IMPULSION.

This negative result was unlooked for, and is at variance with observations announced by Marey. Nevertheless, in view of numerous and varied and carefully conducted experiments, to the end of testing this point, we are compelled to accept the fact as shown and stated.

Obviously the fact teaches that the rate of pulse propagation is not modified directly by the manner of the heart's action; whether it beats quick, launching a sharp wave, or slow, sending a sloping wave, the pulse-wave velocity along the arteries is all the same.

Problem 3.—To determine whether the velocity of liquid waves in the interior of elastic tubes is modified by the size of the tube.

It was first sought to solve this problem by the use of ordinary elastic tubing found in the shops, and so a tube ⅜ inch bore and 6 feet long was put in experiment to obtain results comparable with those of Fig. 25. Fig. 32 was obtained. The time difference is $\frac{1}{12}$ second.

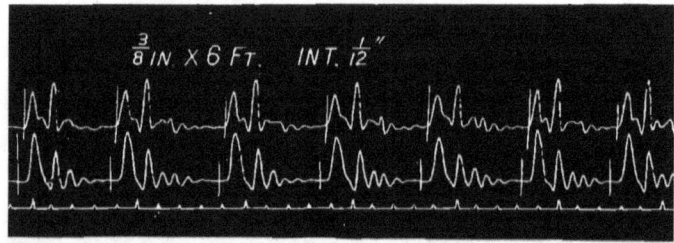

FIG. 32.—TUBE ⅜ INCH, 6 FEET LONG; INT. $\frac{1}{12}$ SEC. COMPARES WITH FIG. 25.

But this experiment and result are not sufficiently conclusive, inasmuch as the larger tube was notably stiffer than the smaller, and this quality, as we have seen, would of itself cause a swifter propagation of the waves. However, as the time intervals in the two tubes were the same, the speeding effect of stiffer walls must have been counterbalanced by a slowing effect of larger size.

To more successfully test the point, it was necessary to experiment with tubes of different diameters, but of the same thinness and elasticity. Accordingly a tube one half inch diameter and two feet long was prepared from the same rubber cloth and in the same manner as the tube $\frac{3}{16}$ inch diameter, which gave Fig. 28. An experiment with this tube gave Fig. 33, in which the time difference is ⅛ second.

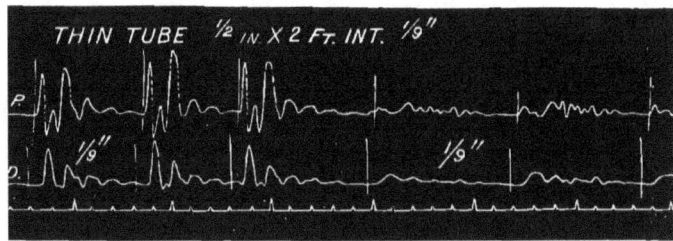

Fig. 33.—Thin Tube ½ Inch Diameter, 2 Feet Long; Int. ⅛ Sec. Compares with Fig. 28.

It will be noticed that the interval in Fig. 28 is ¹⁄₇ second, so this comparative experiment shows that wave propagation is slower in larger and faster in smaller tubes. Compare also Figs. 35 and 36. To the same end Figs. 29 and 30 may also be compared, for, although the gut was laxer than the artery, the fact that the latter gave a slower rate of transmission shows the result was due alone to the larger size.

From this positive result we learn that, other things being equal, the pulse-wave travels slower along larger and faster along small arteries.

Problem 4.—To determine whether the velocity of liquid waves in elastic tubes is modified by a longer or shorter distance from the pump.

For this solution the receivers were placed on a continuation of the ₁₆⁄₃ inch rubber tube, each six feet further from their original positions, and the waves traced at this remoter distance under like conditions as obtained in the production of Fig. 25.

Fig. 34 shows the result—time difference ¹⁄₂ second, the same as in Fig. 25, which represents the nearer distance. However, it appears that the points of the waves are slightly further removed from the beginnings as the distance from the pump increases.

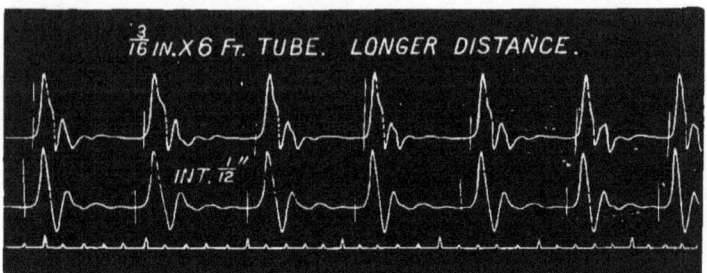

Fig. 34.—Longer Distance from Pump; Int. $\frac{1}{12}$ Sec. Compares with Fig. 25.

We are informed through this experiment that distance from the heart neither accelerates nor retards the velocity of the beginnings of pulse-waves.

Problem 5.—To determine the influence of different pressures on the velocity of liquid waves in elastic tubes.

First Experiment.—Result shown in Fig. 27 where with the softened tube the pressure was successively raised for each wave traced, as shown in the figure. It will be observed that the time difference is the same at 4, 30, 40, and 50 inches pressure, respectively.

Second Experiment.—Result shown in Fig. 29, where with the chicken gut, after the first waves taken at the usual four inches, the pressure was successively raised as indicated. The result here is seen to be positive, the time difference at 4 inches is $\frac{1}{8}$ second, at 30 inches $\frac{1}{15}$ second, at 40 and 50 inches $\frac{1}{18}$ second.

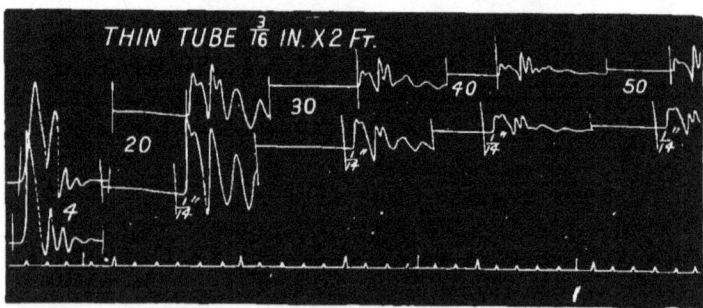

Fig. 35.—3-16 Inch Thin Tube, at Increasing Pressures; Showing Negative Result.

Third Experiment.—Thin rubber tube, $\frac{3}{16}$ inch, at different pressures. Result in Fig. 35.

The showing is negative, the time difference running $\frac{1}{16}$ second throughout.

Fourth Experiment.—Thin $\frac{1}{2}$-inch rubber tube, at different pressures. Fig. 36 gives the result.

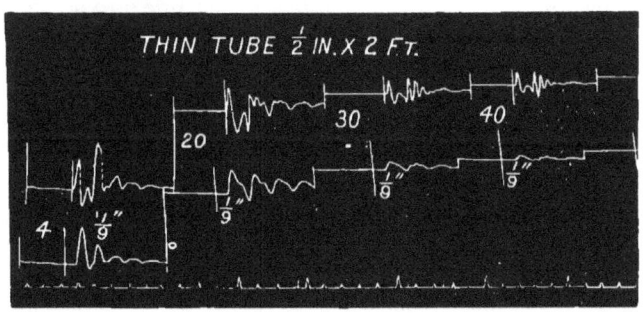

FIG. 36.—1-2 INCH THIN TUBE, AT DIFFERENT PRESSURES; SHOWING NEGATIVE RESULT.

Showing again negative, the time difference measuring uniformly $\frac{1}{9}$ second at 4, 20, and 30 inches pressure, while at 40 inches the interval is really a little longer.

Fifth Experiment.—The calf's aorta, before described, employed and traces taken at different pressures. Fig. 37 shows the result.

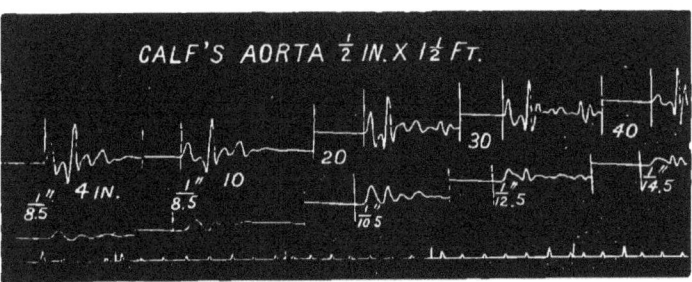

FIG. 37.—CALF'S AORTA, AT DIFFERENT PRESSURES; SHOWING SWIFTER VELOCITY WITH HIGHER PRESSURE.

Intervals respectively $\frac{2}{17}$ second at 4 and 10 inches pressure, $\frac{2}{21}$ second at 20 inches pressure, $\frac{2}{25}$ second at 30 inches pressure, and $\frac{2}{29}$ second at 40 inches pressure. Another positive result.

Reviewing these experiments, it is shown the modifying influence of different pressures is small at best, and requires for development considerable difference of pressure in tubes very soft and elastic. The thin rubber tubes, delicate as they were, failed to make manifest any difference in velocity, while the animal tubes, although with thicker parietes, really more easily yielding, showed increased velocity coincident with marked increase of pressure.

These results have an important relation to the question of influence of blood pressure on pulse-wave velocity. They indicate that variation of blood pressure tends to produce variation of pulse-wave velocity directly as the pressure; but in such pressure changes as occur in the organism and mixture of modifying agencies with which they act, it would be expected, in view of these results, that such effect would be of uncertain manifestation and slight when observed. According to these experiments the current teaching on this point requires modification.

Problem 6.—To determine whether the velocity of liquid waves in elastic tubes is modified by rapidity of current through the tubes.

Hitherto our experiments have been made with the liquid at rest in the tubes, except as sent forwards at each impulsion of the pump. To test the effect of a continuous current on the velocity of waves implanted upon it, the reservoir was elevated thirty-six inches (the supply tube lengthened accordingly), and the distal tube left to discharge freely into a vessel on the table below; and, whilst thus the water was flowing rapidly through the tube, the experiment was made. Fig. 38 gives the result with the $\frac{3}{16}$ inch 6 feet elastic tube. The time difference, $\frac{1}{12}$ second, is the same as that of Fig. 25 given by the same tube with the liquid at rest.

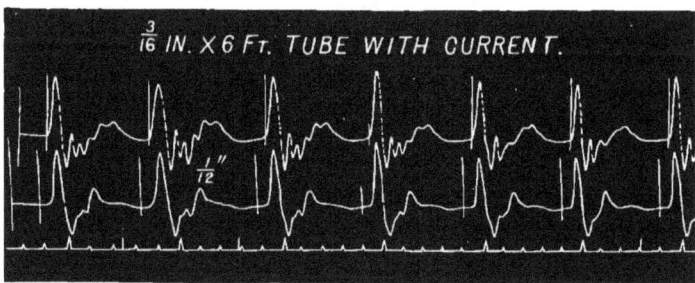

FIG. 38.—3-16 INCH 6 FEET TUBE, WITH A CURRENT; SHOWING NEGATIVE EFFECT.

Fig. 39 is the result of a parallel experiment with the $\frac{3}{16}$ inch 2 feet thin rubber tube. This compares with Fig. 28, given by the same tube with the liquid at rest. It will be observed that the time differences are the same, viz., $\frac{1}{72}$ second.

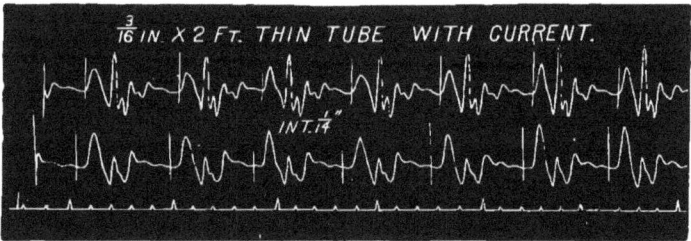

FIG. 39.—THIN 3-16 INCH 2 FEET TUBE, UNDER A CURRENT; SHOWING NEGATIVE RESULT.

Liquid waves, then, travel along elastic tubes with the same speed, whether the liquid be at rest or freely flowing.

By this we may know that, whether the blood in the arteries flows fast or slow, the velocity of the pulse-wave is not affected.

Problem 7.—To determine whether the velocity of liquid waves in elastic tubes is modified by branches issuing therefrom.

Two rubber tubes, each ⅛ inch bore and 6 feet long, were branched on the $\tfrac{3}{16}$ inch 6 feet tube a few inches below the upper receiver, their distal ends turned into the reservoir. Traces of the waves were then taken with the result shown in Fig. 40, which signals a negative effect.

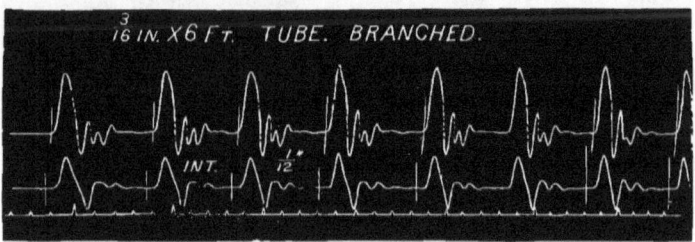

FIG. 40.—SHOWING NEGATIVE EFFECT OF BRANCHES ON MAIN TUBE.

Again, the $\tfrac{3}{16}$ inch 6 feet tube was branched on the chicken's intestine, and Fig. 41 obtained at 10 in. pressure, the first part without, the second part with, communication with the branch. It will be seen that the effect is again negative, the intervals measuring $\tfrac{1}{10}$ second under both conditions.

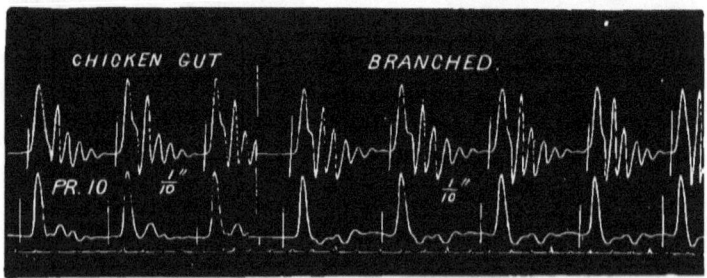

FIG. 41.—CHICKEN GUT, WITH BRANCH; SHOWING NEGATIVE EFFECT.

In application of this experiment, in seeking to determine the cause of different rates of pulse-propagation for different articles, we may exclude as of no effect the different conditions as to branches.

PULSE-WAVE VELOCITY.

Problem 8.—To determine whether the velocity of liquid waves in elastic tubes is modified by the consistence of the liquid.

A solution of boiled starch, as thick as would flow through the tubes, was substituted for water, and experiments made as with water. Fig. 42 shows the result with the $\frac{3}{16}$ inch 6 feet tube, comparable with Fig. 25; and Fig. 43 shows the result with the chicken gut, which is comparable with Fig. 29. Results negative.

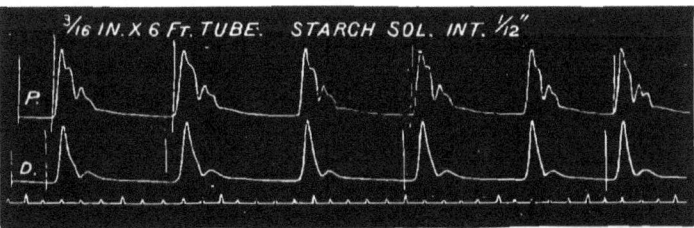

FIG. 42.—$\frac{3}{16}$ INCH 6 FEET TUBE, WITH SOLUTION OF STARCH; NEGATIVE RESULT.

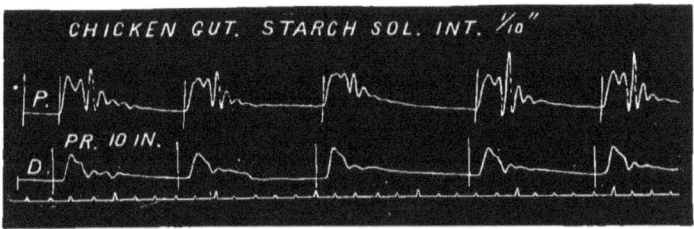

FIG. 43.—CHICKEN GUT, WITH SOLUTION OF STARCH; NEGATIVE RESULT.

This fact teaches that whether the blood be dense or watery, the pulse-wave velocity is all the same.

Problem 9.—To determine the effect of obstruction of the tube on the time of the wave below the obstruction.

In Fig. 44, with the $\frac{3}{16}$ inch thin tube, the waves were traced first under the usual conditions, and then, whilst the tube was compressed just below the upper receiver, and then again just above. The measurements are $\frac{1}{14}$ second with tube free, $\frac{1}{11.3}$ second with obstruction below, and $\frac{1}{14}$ second with obstruction above, the receiver. These fairly represent the results of many similar experiments. The delay is small, but always discernible when the passage of the liquid is greatly obstructed below the near receiver. Obstruction above never causes delay. Also, we found the same result when the obstruction was created by plugging the tube so as to leave a very small aperture for passage of the liquid.

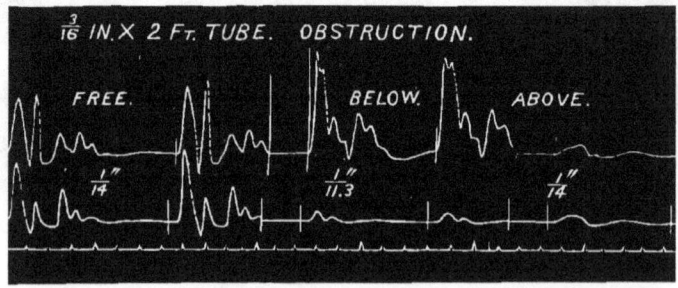

FIG. 44.—$\frac{3}{16}$ INCH THIN TUBE; SHOWING THE EFFECT OF OBSTRUCTION OF THE TUBE.

From this experimentation we learn that waves are delayed by great obstruction of the tube, and that the delay occurs at the point of obstruction, and is not caused by lessening of the rate of transmission below.

The bearing of these facts upon the influence of arterial obstruction in modifying the time of the pulse-wave, is at once apparent.

Problem 10.—To determine the effect of an elastic pouch communicating with the tube, as an aneurism with an artery, on the time of the wave below the pouch.

A thin rubber bag, easily distensible, was placed in relation with the $\frac{3}{16}$ inch 6 feet tube, so that communication between it and the tube could be opened or closed at will. Tracings were then taken, first with the sac shut off, and next with the sac in free communication. Fig. 45 shows the result. The retardation is sufficiently striking.

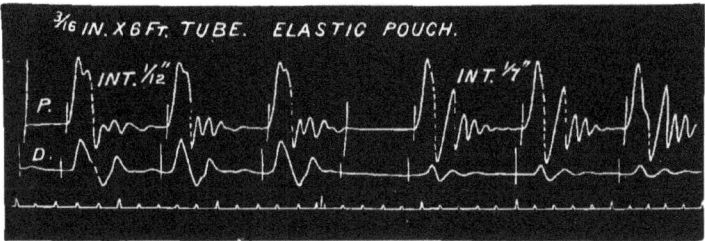

FIG. 45.—$\frac{3}{16}$ INCH 6 FEET TUBE, WITH ELASTIC POUCH.

Next the pouch was placed in the same manner in relation with the chicken gut, and the experiment proceeded with in the same way. Fig. 46 shows the result, which is negative as to delay.

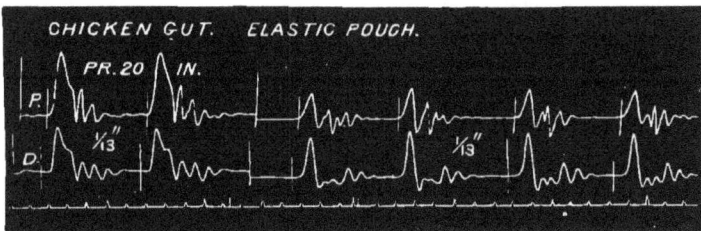

FIG. 46.—CHICKEN GUT, WITH ELASTIC POUCH.

Again, the pouch was associated with the $\frac{3}{16}$ inch thin tube, and in continuity instead of by lateral communication; and to make the experiments strictly comparable a section of the same tubing of the same length of the pouch was interposed for the normal experiment, and that with the pouch above the receiver. First, the pouch was placed closely above the upper receiver and traces obtained shown in the first part of Fig. 47; next, it was removed, and, the original connection having been restored, the middle part was obtained; third, it was placed immediately below the receiver (the section of tube having been removed), and the latter part obtained.

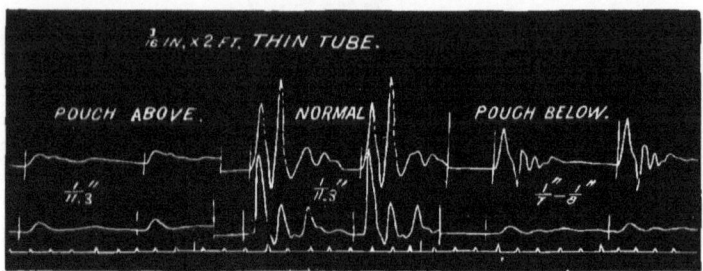

Fig. 47.—Thin $\frac{1}{16}$ Inch Tube, with Elastic Pouch in Different Positions.

It will be noticed that the interval with the sac above the receiver, and that under normal conditions,* without the sac, measure the same; while that with the sac below the receiver is considerably longer.

In the first experiment the pouch was notably more distensible than the tube, and so absorbed the wave to the extent of delaying its time of appearance at the lower receiver; while in the second experiment the contrast in yieldingness between the pouch and gut was not great, and so no retarding effect was contributed.

The third experiment demonstrates that the wave-delay incident to an elastic pouch begins and ends at the pouch, and that the wave which passes through travels on with undiminished velocity. It shows also in connection with the others that the retarding effect is more easily produced by a sac in continuity than by one with lateral connection.

We appreciate the significance of these facts in their bearing upon the subject of delay of the pulse produced by aneurism.

* The longer normal interval than those before shown for the same tube is attributed partly to the length of the tube having been increased by so much as the sac and its connections would increase it, and partly to relaxation of the tube from use and soaking in water.

Having thus experimentally investigated the problems that seemed most pertinent to this branch of our subject, and given somewhat in detail the experiments and results relating thereto, the following resumé of the facts arrived at will now be in order:

1. The velocity of liquid waves along the interior of elastic tubes is proportional directly to the stiffness, inversely to the elasticity, of the tube traversed.

2. It is not sensibly modified by the mode of impulsion, a quick wave and a slow wave being transmitted along the same tube in equal times.

3. It is proportional inversely to the largeness of the tube.

4. It is not sensibly modified by different distances from the pump.

5. It increases with increase of pressure of the liquid in very soft yielding tubes, but in all other elastic tubes it shows no modification.

6. It is not modified by rapidity of current through the tube.

7. It is not modified by branches connected with the main tube.

8. It is not modified by liquids of different consistence.

9. The distal wave is notably delayed by obstruction of the tube, although its velocity of propagation is not appreciably diminished thereby.

10. The distal wave is delayed by communication with an elastic pouch more easily distensible than the tube; while if the pouch and tube are nearly equally yielding, there is no increased delay, yet the velocity of the distal wave is not perceptibly diminished from this cause. Hence,

11. The increased delay of the distal wave in arterial obstruction and distensible pouch arises from arrest at the site of obstruction and site of the yielding pouch.

CHAPTER IV.

PULSE-WAVE VELOCITY.—(*Continued.*)

Facts and New Experiments in Illustration of the Variations of Pulse-Wave Velocity in Man, and Bearing upon the Elucidation of the Causes Which Produce Them.

THIS division of the subject we will study under the form of a series of propositions, and it is intended that the facts brought forward to substantiate the propositions stated shall also enlighten on the causes of the variations in question.

Pulse-Wave Velocity.

Proposition 1.—The velocity of the pulse wave along the arteries increases with increase of age. This proposition has already been proven by our previous researches. We observed and measured the pulse-wave velocity in a child of four and a half years, a man of twenty-five, and another of fifty. In the first the general velocity was 216 inches per

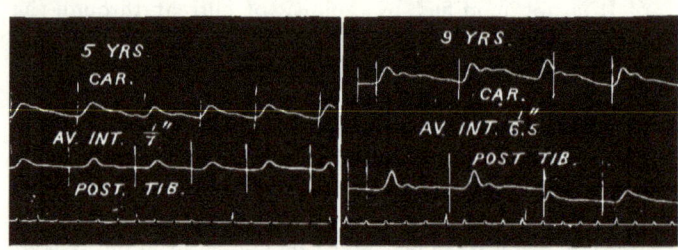

FIG. 48. FIG. 49.

second, in the second 306 inches, and in the third 416 inches per second. As a pertinent illustration of the fact in question, the accompanying figures representing new experiments may be studied. Fig. 48 is from a child five years old, and gives a carotid, posterior tibial interval of one-seventh second; his arterial length between the points of observation measured (approximately) 28 inches; hence his pulse-wave velocity was 196 inches per second.

Fig. 49 is from a boy nine years old; carotid, posterior tibial interval averaging two-thirteenth second, arterial distance 38 inches, velocity of pulse wave 247 inches per second.

Fig. 50 is from a man aged twenty-one years; carotid, posterior tibial interval one-seventh second; arterial distance 50 inches; velocity of pulse-wave 350 inches per second.

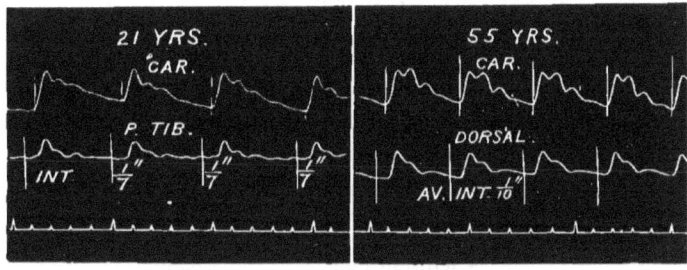

FIG. 50.　　　　FIG. 51.

Fig. 51 is from a man aged fifty-five years; cardio-dorsalis pedis interval averaging one-tenth second; arterial distance 51 inches; velocity of pulse-wave 510 inches per second.

The fact that pulse-wave velocity increases with age, thus so thoroughly established, signals indeed a range of variation the widest that will be noted in all our results. What now is the cause of this marked variation with age?

Relating to the question as between the child and adult, there are four points that claim to be noticed, viz., feebler pulse-waves; smaller arteries; lower blood pressure; and thinner arterial coats, in the child.

In the light of our experiments with tubes (1) feebler waves would have no modifying effect on pulse transmission; (2) smaller arteries would increase the rate; (3) lower blood pressure would tend to diminish the rate; and (4) thinner arterial coats would decidedly slow the velocity of the pulse-wave. Of these factors of modification, it would seem to be just to consider the smaller arteries and lower blood pressure as about counterbalancing each other, and then the thinner arterial coats would be left by the process of exclusion as the efficient cause of the pulse retardation in young children.

As between younger and older adults, it is plain that the only principle which can be invoked in explanation of the difference in pulse-wave velocity, is the increasing stiffening of the arterial walls with the progress of age.

The point needs to be pressed no further; the velocity of the pulse-wave increases with age, in consequence of the progressive stiffening of the arteries as an effect of advancing years.

Proposition 2.—Arteries stiffened by atheromatous and calcareous degeneration give a rapid pulse-wave velocity.

We have before published a case in which the arteries were greatly hardened by degeneration, as found *post mortem*, and in which the carotid, radial interval was measured at $\frac{1}{30}$ second.

Fig. 52 is from a woman aged sixty-five years, who had a basic systolic murmur and a very rigid, knobby, radial artery, that felt like a cord high up in the arm; and who subsequently died from cerebral apoplexy, the result, undoubtedly, of rupture of a degenerated intracranial artery. In these traces the carotid-radial interval measures $\frac{1}{21}$ second.

These cases and intervals demonstrate the rapidity of pulse-wave transmission along pathologically hardened arteries; and, in connection with the foregoing facts, afford ample proof of the following proposition:

Proposition 3.—The velocity of the pulse-wave is directly proportional to the stiffness of the arterial walls.

Proposition 4.—Variation of the pulse-wave velocity as the result of variation of blood pressure is not easily made manifest.

For purposes of comparison the general blood pressure may be reduced by bleeding, by the hot bath, by nitrite of amyl, etc., and is found reduced in adynamia, and increased notably in Bright's disease. We instance here, first, the hot-bath experiment.

Fig. 53 shows tracings of the right carotid and left radial

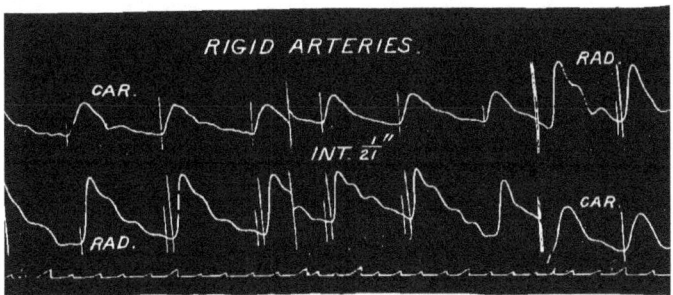

Fig. 53.

pulses from a man aged thirty years, taken just before the bath, and then while in the bath after marked modification in the circulation had been effected.

Notwithstanding the great reduction of the blood pressure, as indicated by the form of the traces and lower pressure, by the manometric tube, at which the radial pulse was best developed, the time intervals before and during the bath were nearly equal, averaging under the first condition about $\frac{1}{12}$ second, and under the last about $\frac{1}{15}$ second.

It was to be expected that such depression of the blood pressure would have signalled a corresponding slowing of the pulse-wave velocity; that it did not was probably owing to the influence of compensating conditions. Thus the arterial tube, in consequence of rapid escape of its contents through the open capillaries into the veins, would diminish the calibre, and, so contracting, thicken its walls. These changes would expedite the transmission of the pulse-wave, and in the case shown more than neutralize the impeding effect of the lowered blood pressure.

The next illustration is from a well-grown boy, aged seventeen years, suffering from severe and protracted typhoid fever complicated with mitral insufficiency. The traces shown (Fig. 54) of the carotid and radial pulses were taken

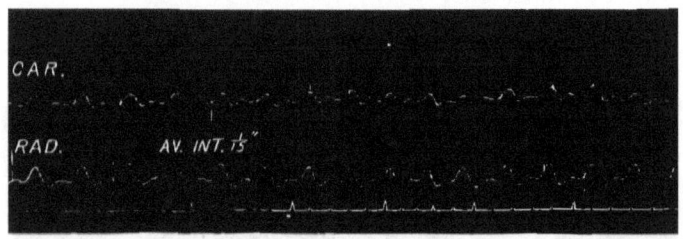

Fig. 54.

on the thirty-sixth day, when adynamia was well-pronounced, temperature 101°, and pulse to the fingers, frequent, small, and very compressible. It will be observed that the average time interval—$\frac{1}{15}$ second—corresponds closely with normal showings, notwithstanding the evident low blood-pressure under which the experiment was made.

The default in this case may be explained as in the last; only the retarding effect of the extremely low blood-pressure would be counterbalanced by the accelerating effect of arterial tubes contracted to adapt themselves to contents lessened by general anæmia as well as free capillary passage, to say nothing of the inadequate arterial supply as an effect of the mitral regurgitation present in the case.

The third illustration is furnished by the influence of nitrite of amyl on pulse-wave transmission.

Fig. 55 is from the same man, who gave the traces (Fig. 53) of the hot-bath experiment. In this instance his left carotid and right radial were taken, it being observed that his right radial was more superficial and gave a better trace than the left. When all was ready his standard was first traced, as shown in the first part of the figure, then the slide was halted, the explorers being kept in their positions and the amyl inhaled until its peculiar effects were manifest, when the slide was started again, and the experiment finished.

The carotid-radial interval measures from $\frac{1}{15}$ to $\frac{1}{13}$ second before, and about $\frac{1}{11}$ second after, the inhalation.

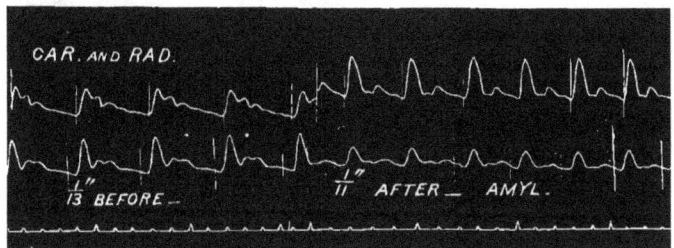

Fig. 55.

In this experiment vaso-motor paralysis would dominate all other modifying factors, joining its retarding force to that of induced lowered blood pressure, and antagonizing reduction of arterial calibre, and so obviating or diminishing the speeding effect that would otherwise ensue therefrom.

Continuing the researches on the influence of different blood-pressures, we will next study the effect of variation produced in certain arteries. The blood pressure in the arteries of a limb is markedly depressed by elevation, and elevated by depression of the limb. Accordingly if the two radials, which normally are synchronous at the same level, are traced with one arm considerably higher than the other, the result will be instructive as to the influence on pulse transmission of a suddenly lowered blood-pressure. Or the same end may be accomplished by tracing a carotid and radial with the radial first depressed and then elevated, and afterward comparing the intervals before and during the elevation of the arm.

The experiment by one or the other method we have performed many times, and usually with more or less retardation of the elevated pulse, but sometimes without any difference.

Figs. 56 and 57, taken from different persons and by the two methods mentioned, exemplify very common phases of result.

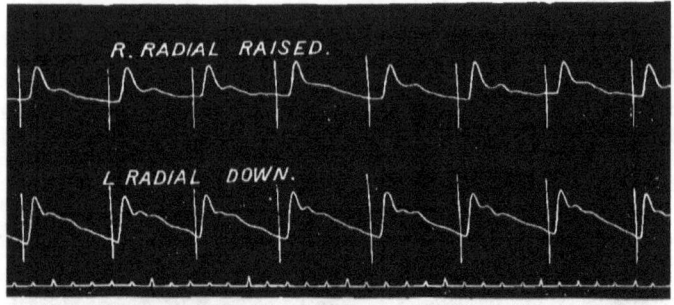

FIG. 56.

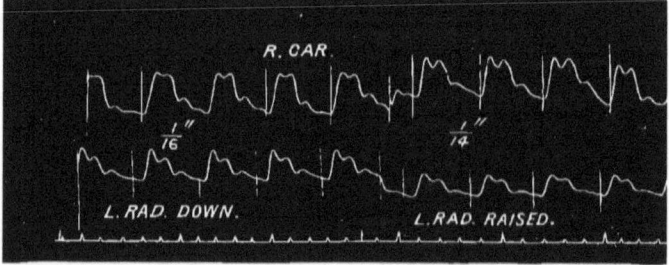

FIG. 57.

The fact that the delay is often small and sometimes fails to manifest under conditions of such revulsion of blood pressure leads us to seek, and, we believe, to find the explanation of the variable effect of the experiment in behavior of the arterial coats. When delay is great the coats are left relaxed after retreat of the blood; on the contrary, when the delay is slight or *nil*, the coats contract as the contents depart.

The change from standing to lying with trunk and head horizontal and lower limbs highly elevated, must cause a very considerable augmentation of blood pressure in the arteries of the upper extremities.

Fig. 58 is an example of the carotid and radial traced under these opposite conditions.

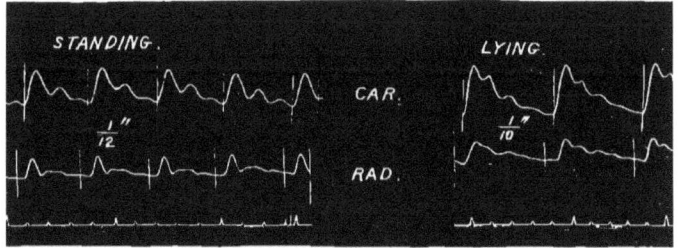

Fig. 58.

The intervals are $\frac{1}{12}$ second standing and $\frac{1}{10}$ second lying. So we have here the paradoxical result of a slower pulse-wave velocity following directly upon increase of blood pressure. Can we explain the phenomenon? The speeding influence of increased pressure is antagonized by the slowing influence of enlarged tubes and walls made thinner and more elastic by distension. We may suppose that arterial tone is a somewhat variable entity, as stimulated by interior pressure; in one instance permitting the fibres to relax to an extent, softening the walls; in another tightening the fibres stiffening the walls.

"The effort," by which is meant making a strong expiratory effort with the glottis closed, compresses the aorta and thoracic and abdominal viscera, driving the blood into the arteries of he extremities, raising their blood pressure in a marked degree. We have often made the experiment for testing the rate of pulse transmission before and during the effort, and almost invariably with the result of proving a swifter transmission during the effort.

Fig. 59 is a fair example of result in this experiment in which the carotid and dorsalis pedis were traced.

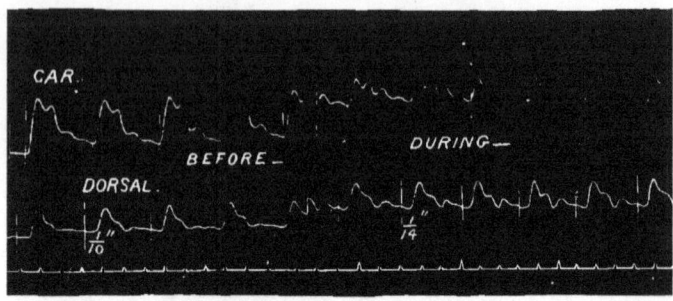

Fig. 59.

Intervals about $\frac{1}{10}$ second before and $\frac{1}{14}$ second during.

Sudden compression of the femorals is another means of augmenting the pressure in the arteries of an upper extremity.

Fig. 60 is from a boy, aged nine years, the same that pro-

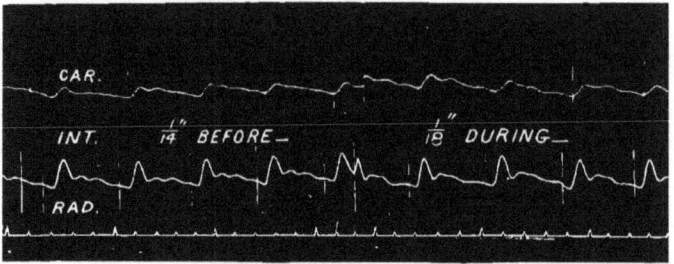

Fig. 60.

duced Fig. 49, taken before and during compression of femorals.

Carotid-radial interval $\frac{1}{14}$ second before and $\frac{1}{18}$ second during.

In explanation of the result in the last two experiments we only remark that the modifying forces were so acting as to throw the balance on the side of pulse-wave acceleration.

It is not deemed necessary to produce other experiments in illustration of the proposition before us. All the experiments go to demonstrate the inefficiency and uncertainty of variation of blood pressure as a direct modifier of pulse-wave velocity. Indeed, in view of the results obtained, it may fairly be questioned whether variation of blood pressure acts at all except as it influences arterial elasticity. We have seen, how, in experiments with inert tubes, increased pressure produced no increase of wave velocity until tubes were employed whose walls were as lax or laxer than those of living arteries. And we have just seen how in experiments on living arteries no certain acceleration follows increase, or retardation decrease, of blood pressure. Such a result both on the schema and man was unexpected, nevertheless the logic of facts must be accepted.

Proposition 5.—The velocity of the pulse-wave varies without notable change of the conditions.

Traces already produced afford abundant evidence of this incessant oscillation of pulse-wave velocity. A critical measurement of the successive time differences in the figures will prove a slight variation between most of them, and between some a difference quite marked. For illustration we refer to Fig. 27, where the carotid-posterior tibial interval in successive pulsations, quite uniform, changes from $\frac{1}{6}$ to $\frac{1}{4}$ second; also we produce a new illustration:

Fig. 61 is from the same young man who furnished Fig.

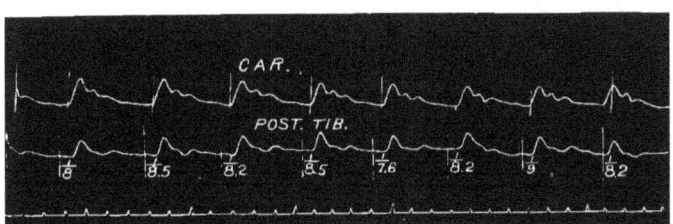

Fig. 61

50, and both were taken on the same day. The time differences of Fig. 50 vary but little from $\frac{1}{4}$ second throughout, while those of the present figure show considerable variation, and the average is less than $\frac{1}{8}$ second. So here are two runs of traces taken from the same arterial points of the same subject, under similar conditions, and the time intervals in one averaging $\frac{1}{7}$, and in the other less than $\frac{1}{8}$ second.

Further we carefully measured each time interval of Fig. 61, and marked the result in the fractions on the slide. The measurements, converted into decimals, read in order, .125, .117, .122, .117, .131, .122, .111, and .117, second; which gives an average of .120 second. Besides, on another slide, with traces taken from the same individual some months previous, the time differences vary around $\frac{1}{8}$ second.

If it be thought that these variations may be results merely of instrumental and mensural errors, we reply that this is impossible. In the method employed it has been proved that the range of instrumental error is so small that it may be neglected, and possible errors of measurement are insignificant, compared with these differences. Again, when we measure successive intervals between waves in inert tubes, the fractions obtained are remarkably uniform. Thus in Fig. 25 the formula of measurement runs without material variation—$\frac{5}{24}$ of $\frac{2}{5} = \frac{1}{12}$ second; and if it be found that the numerator varies from 5, it will also be found that the denominator correspondingly varies from 24.

Another point is, that in the repetitions with the same tube, under the same conditions, the intervals were always the same, while in repetitions with the same individual, under apparently the same conditions, the intervals are often unequal.

Then we deem it sufficiently demonstrated that the variations in question have a real existence.

One theory alone can be offered in explanation of these variations, viz., that the state of contraction of the arterial fibres varies at short intervals, and, so hardening or softening the arterial tubes, causes the pulse-wave to travel with swifter or slower velocity. No other modifying factor can be invoked, and variation of arterial elasticity we have seen is a certain and potent modifier; moreover, there can be no hesitation in accepting as a physiological fact such implied variation of arterial tone. In a word, under the conditions named, the velocity of the pulse-wave varies in consequence of variations of arterial tone; increase of tone causing increase, and decrease of tone, decrease of velocity.

Proposition 6.—The pulse-wave velocity is not modified by variations of pulse frequency. This fact has been sufficiently illustrated by results already produced. Instance Figs. 53 and 54, in which, notwithstanding marked increase of frequency of the pulsations, the time intervals remained at normal values.

Indeed, the proposition would be sustained by considering, first, that the artery starts to rise from the same status at each pulsation, whether the intervals between the waves be long or short; and, second, that quickness of pulse, which usually accompanies frequency, can have no influence upon pulse transmission, since it has been demonstrated that a quick wave and a slow wave travel along the same elastic tube at the same rate of speed.

Proposition 7.—The velocity of the pulse-wave is different for different arterial tracts. This proposition has already been well established, and for its illustration here is produced a figure previously published,[*] which gives the result of an experiment in which the pulsation of the heart, carotid, femoral, radial, and posterior tibial arteries, and the time in fifths of seconds, were traced simultaneously.

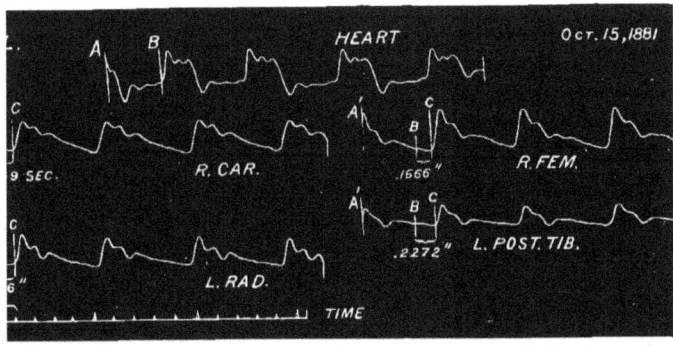

FIG. 62.

[*] *New York Archives of Medicine*, June 1882, p. 231.

This experiment, so prolific in facts, gives the following results regarding the pulse transmissions along different routes: Carotid-femoral time, .0797″, which, with a distance of 18 inches, gives a pulse velocity of 226 inches per second. Carotid-radial time, .0797″, which, with 23 inches distance, gives a pulse velocity of 288 inches per second. Femoral-posterior tibial time, .0606″, which, with 33 inches distance, gives a pulse velocity of 544 inches per second.*

Thus it is well shown how the rate of pulse transmission varies in value along these different lines, and now the problem as to the cause of these differences presents for solution. Pertaining to the arteries there are four points of difference in condition, viz.: (1) difference as to giving off of branches; (2) distance from the heart; (3) state of elasticity; and (4) size. The question as to any modifying influence of the first two has been settled in the negative by our results with tubes. As to the third, it is a well-known fact that the aorta is highly elastic; its coats are thick, yet extremely soft and pliable, and yield with the greatest readiness to increase of inferior pressure, to promptly return when the pressure diminishes. The aorta, it is safe to say, is more exquisitely elastic than the arterial trunks of the extremities, and to this difference in elasticity we are led to attribute in part the comparative slowness of aortic pulse transmission. And in regard to the small but real difference of rate between the upper and lower extremities, an assumed greater resistance in the arteries of the lower is the only explanation that offers of the faster transmission in the latter.

Difference in size of the arteries is in reality the potent factor which makes the pulse travel slower along the aorta than along the other arteries. This conclusion is inevitable when we remember the law that wave-velocity is proportional inversely to the size of the tube traversed, and the

---

* The cardio-arterial intervals are expressed on the cut, and the arterial intervals are obtained by deducting the lesser cardio-arterial from the greater; thus, cardio-femoral interval .1666″ — cardio-carotid .0869″, gives .0797″ as the carotid-femoral interval, and so on for the other arterial intervals.

fact that the aorta is many times larger than the arteries of the limbs.

Proposition 8.—The time of appearance of the pulse-wave is delayed in arteries in which the blood column has been much reduced by obstruction of the current.

The experiment upon which this proposition depends for proof is readily made on man by tracing the two radials before and during compression of one axillary or brachial; or by tracing the two posterior tibials or dorsals before and during compression of one femoral or popliteal; or again, the method may be pursued of tracing a near and more distant arterial point, as the carotid and radial, before and during compression of an intermediate point, as the axillary or brachial.

Figs. 63 and 64 are examples of results from such ex-

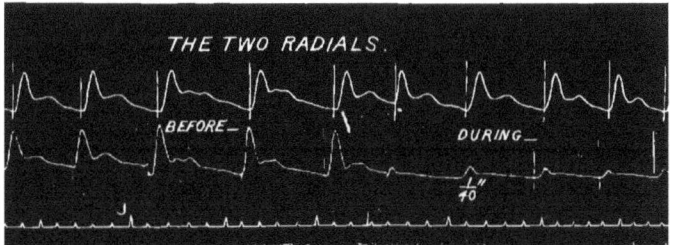

Fig. 63.

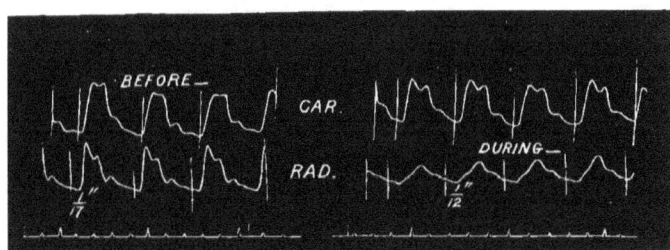

Fig. 64.

periments on two individuals of different ages. The delay in either occasioned by the obstruction is about $\frac{1}{40}$ second.

In explanation of pulse delay from this cause, from our experiments on tubes, we need take no other account than that of check at the site of obstruction. The condition of the artery below, if contracted and diminished in size (which is probably the case), would only tend to lessen the delay.

Proposition 9.—The velocity of the pulse-wave is diminished in arterial trunks affected with vaso-motor paralysis.

This proposition is a necessary corollary of facts already acquired. In vaso-motor paralysis the arterial coats are relaxed and the arterial calibre enlarged, both of which conditions are effective factors of pulse retardation. Besides, we have published a case with tracings in which the phenomena of general vaso-motor paralysis were well declared, with coincident remarkably slow pulse transmission.

Proposition 10.—The time of appearance of the distal pulse is delayed in aneurisms with yielding walls and free communication with the artery; while there is no delay in aneurisms with resisting walls, even though freely communicating, unless the aneurisms directly obstruct the arterial current, or diminish wave velocity by producing vaso-motor paralysis.

The clinical proofs of this compound proposition have been well furnished by François Franck[*] and the author,[†] whose publications contain graphic illustrations of the points from actual cases.

From this study we deduce the following resumé of the more prominent facts:

1. The velocity of the pulse-wave is determined above all by inherent states of arterial elasticity; being slower as the arteries are more elastic.

2. It is incessantly changing, within small limits, in consequence of variation of arterial tone; being faster as the tone is higher.

3. It diminishes with the size of the artery traversed.

[*] *Jour. de l'Anat. et de la Physiol.*, t. xiv. (Mars–Avril, 1878), et t. xv. (1879).
[†] *N. Y. Med. Record*, Nov. 29, 1879.

4. It tends to increase with increase of arterial pressure, but modification from variation of pressure often fails to manifest.

5. It is not perceptibly modified below the site of an arterial obstruction, but the distal wave is delayed there in consequence of check at the site of obstruction.

6. It is not perceptibly modified in an artery below the site of an aneurism, although the distal wave may be delayed there in consequence of absorption by the yielding aneurismal walls.

CHAPTER V.

THE PRESPHYGMIC INTERVAL.

The Causes of the Variations of the Cardio-Aortic or "Presphygmic" Interval.

ONE of the outcomes of the simultaneous graphic method has been the demonstration and measurement of an interval of time between the beginning of ventricular contraction of systole and opening of the aortic valves, or beginning of the aortic pulse. The term " presphygmic," applied to this interval by the author, appears well chosen, as expressing the phase of ventricular systole which precedes the rise of the arterial pulse. This term and " cardio-aortic" or " ventriculo-aortic," will be used synonymously.

Observations relating to the presphygmic interval must necessarily be made on man between the point of ventricular beat and the point of pulsation of a near artery, as the carotid or subclavian; the former being usually selected as more accessible for exploration. To get the presphygmic interval in its purity, it is only required to deduct from the full cardio-carotid interval the brief transmission interval of the pulse-wave between the points named; but so short is this interval compared with the whole cardio-carotid time difference that practically it may be neglected, and the time between the ventricular beat and carotid pulse taken as the representative of the presphygmic interval

Experiments on the schema in aid of elucidation of questions pertaining to the presphygmic interval must be made between a pump representing the heart and near point of a tube representing the aorta. Accordingly, for our purpose, the upper receiver was placed in communication with the interior of the pump of the schema, and the lower receiver with the interior of the egress tube, twelve inches distant. Thus, compression of the pump or ventricle immediately increases the pressure within, which increase sooner or later overcomes the valvular barrier, and is felt in the arterial tube. The two events of ventricular and arterial increase of pressure are instantly signalled as waves, and their time-relation to each other is then easily determined. The arterial tube, being practically rigid, would give for the short distance traversed an inappreciable transmission interval, so all the delay signalled by the traces may be placed to the account of the schematic presphygmic interval. The action of the hand on the pump can be made to imitate very closely the movements of the human ventricle.

In the present study the form of distinct propositions will be continued.

Proposition 1.—The duration of the presphygmic interval varies with the pulse rate; being shorter with frequent and longer with rare pulsations.

Experiment on man never fails to prove variation, in the sense stated, of the cardio-carotid interval coincident with a considerable variation of pulse rate. We offer two illustrations. In Fig. 65 the heart and carotid were traced before and immediately after active exertion. The pulse rate before was 82, after 130. The cardio-carotid interval was before .075", after .047", all as shown.

Fig. 66 was taken from a girl, aged twelve years, on the second day of scarlet-fever. Temperature 103.2°, pulse rate 126. The tracings are of the heart and radial, and carotid and radial. The cardio-radial interval measures $\frac{1}{10}''$ and the carotid-radial $\frac{1}{15}''$, which would make the cardio-carotid interval only $\frac{1}{30}''$. We could easily multiply such examples.

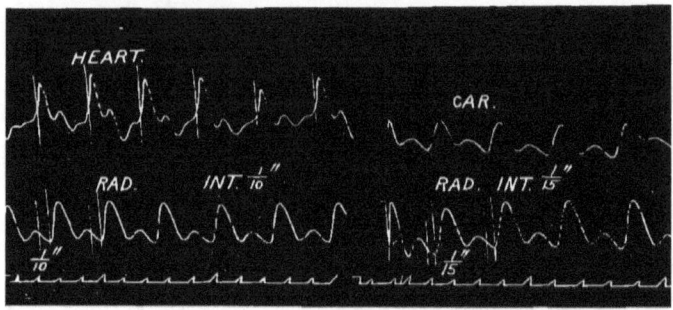

Fig. 66.

Coincidence of cardio-carotid lengthening and pulse infrequency has its limits. We have found that an interval of $\frac{1}{10}''$ usually goes with a pulse of 60, but if the latter fall lower, there is no certainty that the former will farther lengthen.

This relationship between the pulse rate and cardio-carotid interval for variations between 60 and 120, the author expresses by the following working formula: The cardio-carotid interval is normally about one tenth the duration of the pertaining pulsation. Thus a pulse of 60, 1 second long, would give for the interval $\frac{1}{10}$ second; a pulse of 72, $\frac{5}{6}$ second long, would give $\frac{1}{12}$ second, and so on. Any considerable departure from this ratio, we consider, would be irregular, and dependent upon abnormal conditions. New observations confirm in the main the justness of this formula.

Nevertheless pulse rate in itself can have no modifying influence upon the presphygmic interval; other things being equal, the rate may be fast or slow and the interval remain the same. With the schema, the ventricle worked at a uniform quickness and force, and the outflow and pressure maintained at a given rate and value, the interval between the rises of pressure in the ventricle and artery will be the same, whatever the order of succession of the impulses.

But in the organism, when the pulse-rate changes, other conditions change likewise, and in these concomitants we shall find the real producers of the presphygmic variations found associated with modifications of pulse rate.

Proposition 2.—The duration of the presphygmic interval varies with the mode of ventricular systole; being longer with slow and shorter with quick contractions.

It would appear that effect must follow cause as implied in the proposition; for a quick action of the ventricle must raise the ventricular pressure, so as to overcome the arterial pressure and send forth the wave sooner than a slow action. But the demonstration is easily made on the schema.

In Fig. 67 the ventricle was first made to contract slowly, and then quickly, with a ventricular pressure of 4 inches, and arterial gradually rising from 50 inches. It is seen that the interval between the ventricular and arterial waves is very much longer under the slow impulsion than that under the quick; the former measuring $\tfrac{1}{5 \cdot 6}''$, and the latter $\tfrac{1}{17}''$.

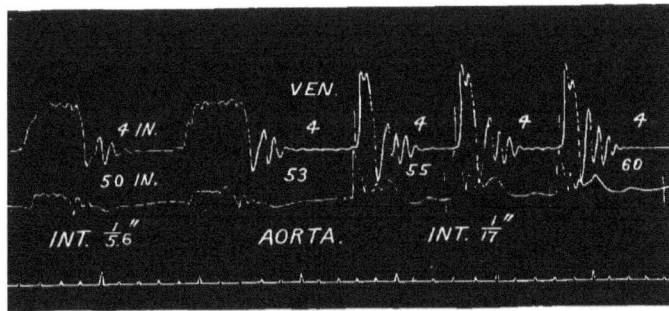

Fig. 67.

It is plain that the greater the difference between the pressure in the ventricle and artery, the greater will be the modifying effect of different modes of ventricular action;

and if the pressures are in equilibrium, a slow action will start the arterial wave as soon as a quick action.

When we seek in man examples of the effect of different modes of ventricular action, we find them in modifications of pulse frequency. The frequent pulse is sent forth by a comparatively quick systole, and the rare pulse by a comparatively slow systole. Figs. 65 and 66, already produced, will illustrate this proposition. It cannot be doubted that a quick ventricular systole characterized the acceleration of movement shown.

It may be true that the arterial pulse is sometimes quick and rare, or slow and frequent; but ventricular systole is probably always slow when rare, or quick when frequent. We can understand such coincidence of slow systole and quick pulse, or quick systole and slow pulse, inasmuch as the quality of the arterial pulse depends upon the arterial as well as the cardiac conditions.

Proposition 3.—The duration of the presphygmic interval varies with the excess of arterial over ventricular blood pressure, and is longer with a high and shorter with a low value of such difference.

Assent to this proposition is readily gained through *a priori* processes. At the beginning of systole, the higher the arterial pressure relatively to the ventricular, the longer must be the time required to raise the latter above the former. One condition alone could defeat such an order, viz.: a quicker initial ventricular contraction coincident with the higher arterial pressure; but the proofs are convincing that the reverse obtains.

In demonstration we offer an example from experiments on the schema.

Fig. 68 shows traces of the ventricular and arterial waves at different degrees of arterial pressure, the ventricular remaining throughout at a uniform pressure of 4 inches. The first waves, with pressure in equilibrium, show no appreciable arterial delay; the second waves, with arterial pressure at 40 inches, show a delay of .08 second; the third waves,

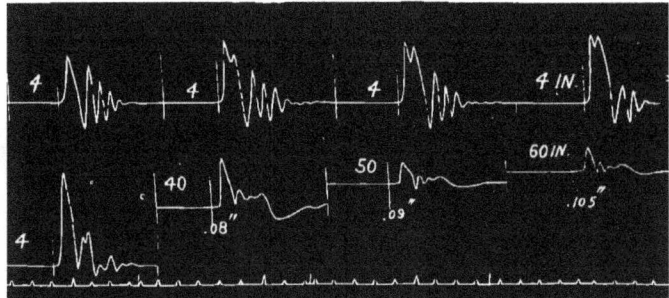

Fig. 68.

with arterial pressure at 50 inches, show a delay of .09 second; and the fourth waves, with arterial pressure at 60 inches, show a delay of .105 second.

In the man experiment of tracing the heart and carotid before and during compression of the femorals, succeeds in showing elongation of the presphygmic interval as a result of increased aortic pressure. Fig. 69 is an example taken from the nine-year-old boy that furnished Fig. 49. The cardio-carotid interval is $\frac{1}{12}$ second before, and $\frac{1}{10}$ second during, the compression. This experiment commends itself for its purity in that there are no complicating conditions.

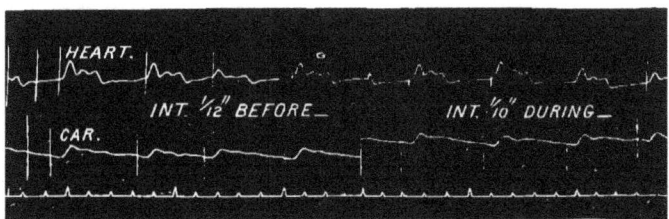

Fig. 69.

Again, in fever the arterial pressure is notably low, and always in this condition, if the heart valves are intact, the cardio-carotid interval is shown to be diminished. In illustration, besides Fig. 66 before given, Fig. 70 may also be studied. It was taken from a boy five years old in the height of scarlet-fever, of which he died two days afterwards. The traces are from the heart and radial, but from these it is easy to approximate the interval between the heart and carotid.

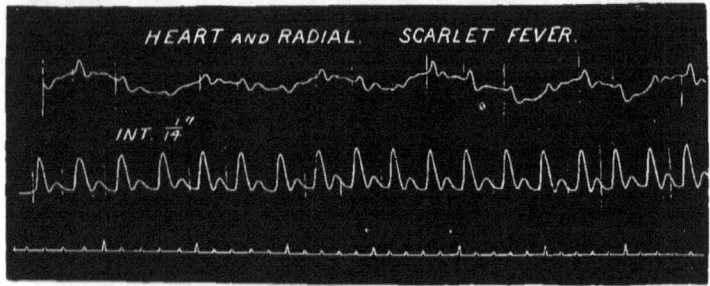

Fig. 70.

The cardio-*radial* interval averages about $\frac{1}{14}$ second, which, even with a rapid transmission time, would make the cardio-carotid interval extremely short; in no event could it exceed $\frac{1}{30}$ second. To make the showing stronger we introduce Fig. 71, taken from a healthy little boy of the same age, on the same day, with the instrument unchanged. It will be observed that the cardio-*radial* interval here averages about $\frac{1}{7}$ second, twice as long as in the former instance.

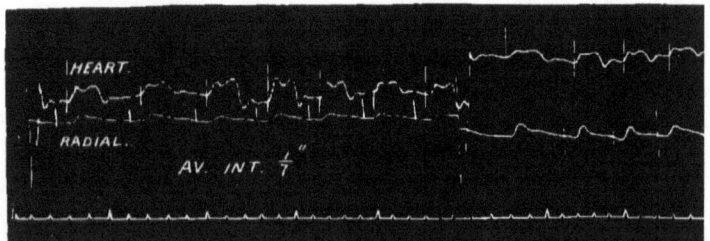

Fig. 71.

However, examples from fever cases are complicated with the effect of quickened systolic contraction; and indeed it is difficult from any source to obtain the effect of lowered arterial pressure disassociated from that of systolic quickening. Nevertheless, it must be true that quickness of ventricular contraction could not produce such shortening of the presphygmic interval without concurrence of the effect of low arterial pressure.

We are now prepared to understand why the duration of the presphygmic interval is proportional inversely to the pulse rate. It is because the pulse rate stands in a sense as the exponent of the mode of systole and relative arterial pressure. When the pulse is frequent, systole is quick and the pressure is low, and the interval is short. On the other hand, when the pulse is rare, systole is slow, and the pressure is high, and the interval is long.

It is worthy of remark that the ventricular and arterial blood pressures, while readily changing their relative value, tend promptly to return to the normal difference, and in the processes of these fluctuations the operations of the modifying factors may again be farther modified. Thus, if the capillaries become constricted, the increased arterial pressure will add itself to the slow ventricular contraction, and the two will produce a marked lengthening of the presphygmic interval; but soon the ventricular pressure rises and the arterial declines, the balance is restored, and the mode of systole and the presphygmic interval again become normal.

Or, again, if the capillaries become suddenly relaxed, the arterial pressure falls, the heart starts off with quickened and accelerated action, and the two factors here unite to produce a marked shortening of the presphygmic interval; but under the increased frequency of the systoles the pressure soon rises again, and whilst rising, the pulse acceleration not yet checked, the factors antagonize each other, and the presphygmic interval may not be diminished though the pulse is frequent. Soon, however, all is regular again. These variations of conditions and effects can be well illustrated on the schema.

Proposition 4.—The duration of the presphygmic interval is subject to limited variation, even when the cardiac action and blood pressure appear most regular and equal.

In illustration of this proposition we will study Fig. 72, selected for the distinct markings and apparent regularity of the pulsations. We took the pains to measure on the slide the cardio-carotid interval of each pulsation, marking the result below each basal point of the carotid traces; also the duration of each cardiac systole and cycle, marking them respectively as shown in the cut. These measurements were made under a glass, with extreme care, and it is believed they contain no material error.

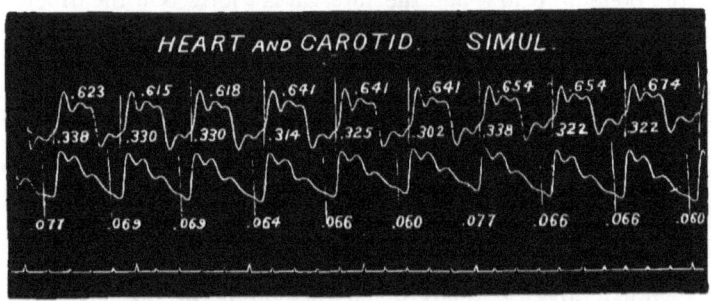

Fig. 72.

The lower row of decimals shows the variations of the cardio-carotid interval. The longest interval is .077" and the shortest .060".

The upper row of decimals shows the duration of the cardiac cycles. Examining the relation between these and the corresponding cardio-carotid intervals, the variations are found discrepant ; a long interval going, as likely, with a short cycle, and a short interval with a long cycle. So in the small variations of cardiac rhythm, the rule does not hold of a ratio between the presphygmic interval and the pulse rate.

The middle row of decimals shows the duration of the cardiac systoles. Examining the relation between these and the cardio-carotid intervals, the variations are found in the same direction ; a long interval going with a long systole, and a short interval with a short systole. The correspondence here shown we have found to hold as a rule in the small variations of systolic duration of so-called regular pulsations. But the rule has exceptions, for the length of systole is determined by its ending as well as by its mode of beginning.

Therefore we feel justified in formulating these observations under the following statement :

Proposition 5.—In the small variations of apparently regular pulsations the presphygmic interval does not observe any rule of variation with respect to the duration of the cardiac cycles, but, as a rule, varies in the same sense as the duration of the cardiac systoles.

In explanation of the proposition we remark there are no facts anywhere to indicate that diastole is any thing but a cipher in the processes ; all depends upon systole. The explanation is found in the fact before developed and applied, viz. : that when the systoles are longer their beginning is slower, which determines a longer interval ; and when the systoles are shorter their beginning is quicker, which determines a shorter interval.

CHAPTER VI.

INFLUENCE OF MUSCULAR EXERCISE ON THE ARTERIAL AND CARDIAC PULSATIONS—THE PULSATIONS OF THE FONTANEL: THEIR FORM AND MECHANISM, AND RELATION TO THE PULSATIONS OF THE HEART AND ARTERIES, AND THE MOVEMENTS OF RESPIRATION.

BEFORE entering upon the study of the pulsations in disease it is important to further extend our knowledge of the physiology of the pulsations. Accordingly we now inquire, What variations in the action of the heart and arteries and related phenomena are produced by active muscular exercise? Having given the subject careful consideration, the results and conclusions may be stated in summary, attesting the more important by illustrations of actual experimental results. It will be understood that the effects under exercise are considered in comparison with the effects under rest.

1. *The effects of exercise on the arterial pulsations.*—The *frequency* is increased, as all know. The *regularity* is undisturbed. The *amplitude* is often increased, but sometimes unchanged, other times diminished. The *form*, too, as regards the secondary waves presents variations; a very common departure is diminished prominence of the second wave. Usually after walking the trace of the pedal arteries shows a relatively higher second wave. The *pressure degree* is in most instances but little changed, at times it is higher, more rarely lower. The *tension*, as indicated by the pressure degree and relative height of the second wave and aortic notch, is little modified from that under conditions of rest;

when changed, however, it is oftener towards a higher degree. *The cardiac systolic portion of the pulse trace compared with the cardiac diastolic portion is lengthened;* thus while under rest with a pulse-frequency of 76 to the minute, these divisions may be to each other as 1 : 2 ; under exercise with a pulse-frequency of 100, the ratio may be as 2 : 3. *The velocity of the pulse-wave shows little departure from the standard of the individual under rest ; and this agreement pertains not only to entire arterial lines, but also to their divisions.*

Figs. 73, 75, with Fig. 74, may be studied so far as their showings apply in exemplification of the above facts and statements. Figs. 73 and 75 represent K's carotid and dor-

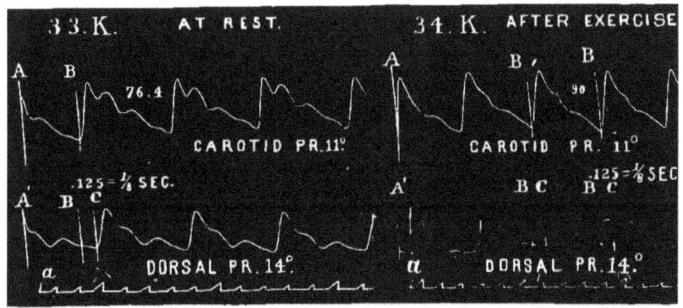

FIG. 73.

salis pedis pulse under conditions of rest and active exercise respectively. The first was taken while he was perfectly quiet, then he started out for a brisk walk, and immediately on returning the second was taken. It will be noticed that the time difference under these opposite conditions measures the same, namely, .125 second, and that this corresponds with Fig. 74, which gives the result of a former experiment on K when at rest, but exhibiting considerable acceleration of the pulse.

2. *The effects of exercise on the cardiac pulsations.* As to the cardiograms, exercise does not always increase their amplitude, or is it usually favorable to their finer delineation.

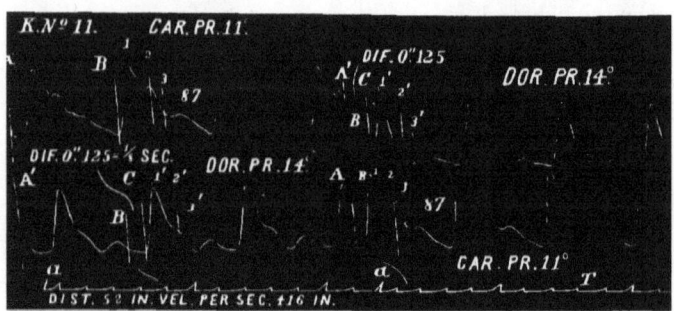

FIG. 74.

Exercise lengthens relatively the duration of ventricular systole. Exercise abbreviates the ventricular presphygmic interval. This is deduced from two demonstrated facts: (*a*) the ventricular-carotid time is shortened, while (*b*) the pulse-wave velocity remains unchanged.

In exemplification of the facts as stated, Fig. 75 may be studied in comparison with Fig. 76, the latter under rest, the former after running up and down stairs. The lessons are easily read. Shortening of the ventricular-carotid time under exercise is clearly demonstrated.

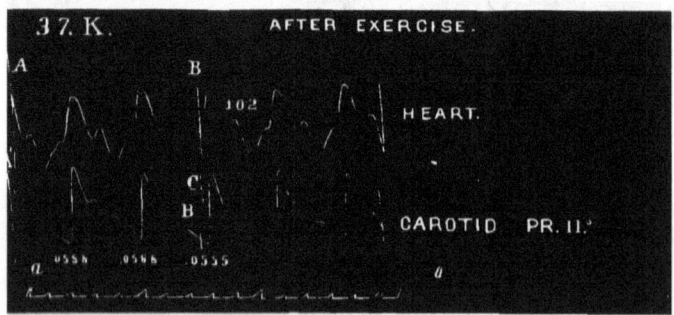

FIG. 75.

Deductions.—From these and data already contributed we deduce:

1. *That the velocity of the pulse-wave does not notably change*

INFLUENCE OF MUSCULAR EXERCISE.   133

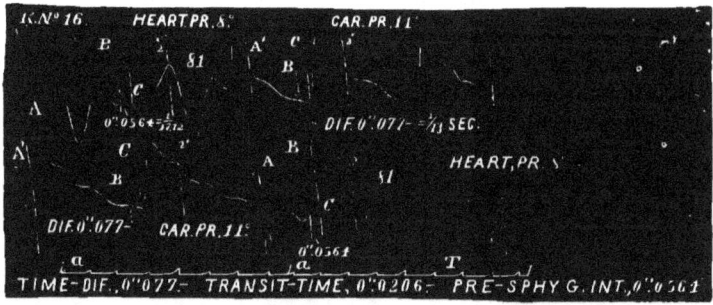

FIG. 76.

with changes of pulse-rhythm;—a rare and frequent pulse, when other conditions are equal, being transmitted over the same arterial spaces in equal time.

2. *That the velocity of the pulse-wave normal to the individual is closely preserved under ordinary physiological variations of the circulation.*

3. *That the presphygmic interval changes with changes of rhythm; the period being longer with rare, and shorter with frequent pulsations.*

The influence of *other conditions* on the pulse-wave velocity and presphygmic interval receives from this exposition no definite elucidation, for the reason that in the experiments arterial tone, blood pressure, etc., exhibited no marked departures from the standards of the individual under normal conditions of rest. However, the operation of these agencies can be studied by more direct experiment; and the solution will be simplified in that we now know the effect of variation of rhythm is negative on pulse-wave velocity, and that pulse-acceleration abbreviates the presphygmic interval.

Observations on Bertha Von Hillern's Pulse under Exercise.

At intervals of rest during the walk of Miss Von Hillern, in Cincinnati, of 100 miles in 28 hours without sleep, two sphygmometric and sphygmographic observations of her left radial artery were permitted. It was impossible to procure a record of the pulse in her quiet before the walk, and to supply, in some degree, this deficiency, the pulse of another person is presented in this connection, under conditions of quiet, known to be one of good amplitude, tension, and force, though usually accelerated when placed under experiment.

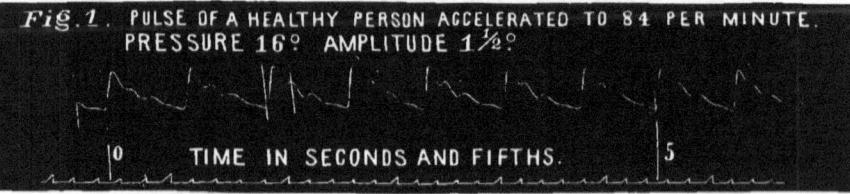

Fig. 77.

The first observation of Miss Von Hillern's pulse was taken fifteen minutes after the completion of her seventy-third mile. See Fig. 78.

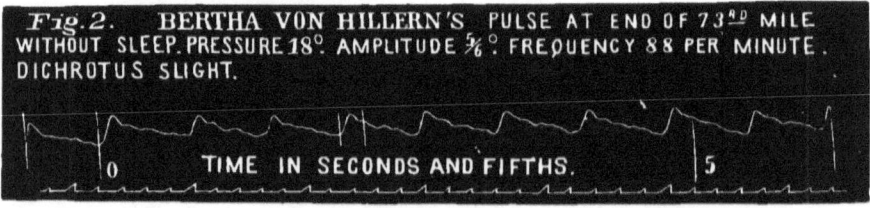

Fig. 78.

The second observation was taken ten minutes after the completion of her ninety-eighth mile. See Fig. 79.

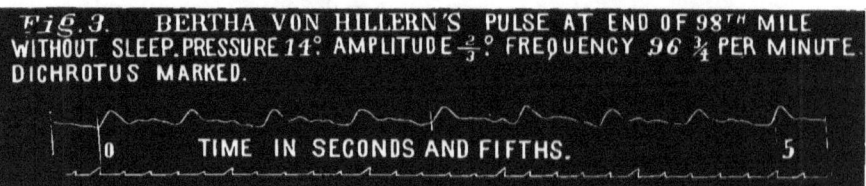

Fig. 79.

The Reading of the Records.—In Fig. 77 the *pressure* is marked 16°, which shows the height in the graduated glass tube at which the greatest undulations were displayed, and indicates the pressure sustained by the artery to develop its best pulsations. The tracings were taken at this same pressure. The *amplitude* is marked $1\frac{1}{2}°$, and indicates that the undulations in the tube passed through this extent of sweep. The *tracing* so far accords with these indications and is easily read, showing the pulsations of good amplitude, with tension rather above than below the average, regular, equal, and running just seven to five seconds of the chronogram, making the frequency 84 per minute. They rise with sufficient suddenness to the summit and then gradually descend by the well-marked steps of the secondary waves, to the basal line.

In Fig. 78 the spirit index marks a pressure of 18° and an amplitude of five sixths of a degree, with very slight dicrotus. The tracing is of moderate amplitude, sloping ascent, shows a well marked proportionately high second wave, also a proportionately high aortic notch. One pulsation, the second, is notably longer than the others, which latter are all nearly equal in rhythm and amplitude. There are seven and a third pulsations to five seconds, which places the frequency at 88 per minute.

The indications of this record appear to point to a pulse of rather small volume, increased arterial tone, and high blood-pressure, combined with considerable vigor of cardiac action. The cardiac energy is demonstrated especially by the distinctness, relative height, and horizontal distance from basal point of ascent of the second or so-called tidal wave, resisted as the heart inevitably was by the high blood-pressure and increased arterial contraction. In Fig. 79 the pressure is 14° and amplitude two thirds of a degree, with marked dicrotus. The tracing is small, sloping, without mark of second wave and distinctly dicrotic; pulsations nearly uniform in rhythm and amplitude, run eight and one sixteenth to five seconds, yielding $96\frac{3}{4}$ to the minute.

The indications of this record point unmistakably to a pulse of diminished volume, diminished arterial tone, and diminished blood-pressure, combined with reduced action of the heart. The reduced cardiac action is demonstrated especially by the low altitude and sloping ascent of the first wave, when the yielding arterial coats and lowered blood-pressure would with certainty insure a quick and free sweep of this wave under the impetus of even very moderate cardiac vigor.

It is not always easy to distinguish between a high-tension *contracted* pulse and a high-tension *expanded* pulse; as both afford tracings of small amplitude, definite secondary waves, and are developed under high pressure of the instrument. Other things being equal it will appear that the expanded high-pressure pulse, as compared with the contracted, will yield a tracing of greater amplitude, quicker ascent, and sharper delineation of the secondary waves; and for the reason that the expanded arterial coats permit their fine elastic play and are more sensitive and responsive to the impetus from the heart. Judged by all the tests the tracing of Fig. 78 is an example of a contracted high-tension pulse. There can be no question that the tracing of Fig. 79 is an example of small low-tension pulse. The contrast is striking and significant. It will be observed that in both tests the heart and arterial system were in unison. In the first, arterial contraction was found coincident with vigorous action of the ventricle; in the second, arterial relaxation coincided with reduced ventricular action.

Observations upon Daniel O'Leary.

During Mr. Weston's five-hundred-mile walk in England, Dr. Mahomed made a series of sphygmographic observations of his radial pulse. The published results (see *Lancet*, Am. Ed., Oct., 1876, p. 479) show a pulse of large amplitude, average tension—the tension slightly increasing during the progress of the walk,—without marked dicrotism, fairly regular, and in frequency from 98 to 134 per minute. Dr. Mahomed attempted, but failed, to get cardiographic tracings.

Observations, made by the author, of Miss Bertha von Hillern's radial pulse while engaged in her late walk in Cincinnati of 100 miles in 28 hours, showed increased tension and diminished amplitude, and a frequency of 88 at the 73d mile, and diminished tension and amplitude, and a frequency of $96\frac{3}{4}$ at the 98th mile.

Mr. Daniel O'Leary, the champion pedestrian of the world, on the occasion of his recent walk at Price's Hill, Cincinnati, of 220 miles in 56 hours, kindly allowed tracings to be taken of his pulsations. Unfortunately Mr. O'Leary's time was so occupied previous to the walk, that he could only spare a few moments in which to obtain his standard tracings, and these immediately before and in the midst of the flurry of preparation for the start. Under the disadvantages, the tracings were imperfect, but they showed, however, the same type of form as those obtained during and after the walk. By this examination his pulse was 76, regular, moderately ample, and dicrotic. During the contest at each observation his pulse was also counted by a watch immediately after his leaving the track, and its frequency was only a few beats more per minute than that of the pulsations recorded and timed by the chronograph about five minutes later. His respiration throughout was singularly unaffected.

The observations were conducted according to the usual method, by means of the special recording apparatus, of tracing simultaneously the pulsations of the heart and an artery, or of two arteries, and a chronogram, in seconds and fifths, all on the same slide. In this way, with the assistance of Dr. H. T. Lowry, without trouble or delay the tracings which are shown in the illustrations were secured. Like records were also obtained of Mr. O'Leary's heart and carotid, carotid and radial, carotid and femoral, radial and dorsal, and femoral and dorsal.

FIG. 81.

FIG. 82.

The records are wholly trustworthy. The instrument was in perfect order, and at the standard adjustments; moreover, thorough testing before and after the experiments proved its action perfect and uniform.

The sweep of a cardiogram in a given case is governed by other conditions besides the force of the heart's contractions. Of two hearts beating with equal vigor, one behind a thin chest-wall and wide intercostal space, the other behind a thick chest-wall and narrow intercostal space, the first will yield a high-sweep and the second a low-sweep tracing. The effect of position, too, is very marked. If the person is lying flat on the back there will be little, if any, result; if standing or sitting erect, the result will be plainer; if leaning forward, it will be still better; but if lying well over on the left side, the sweep will be the highest of all. So the systolic vigor may be judged by the amplitude of the tracing only when all the attending circumstances are taken into the account. Mr. O'Leary possesses a compact and muscular chest-wall. The fact that No. 2 cardiogram is less high than the others, is the result of position, in that he was lying more on his back whilst it was being traced. The irregular line of No. 4 cardiogram is the result of the disturbing effect of the respiratory movements, which were then, really, more pronounced than at the periods of observation during the walk. In this cut the well-traced pulsations are at either end and the middle; the others being defective from the cause named. The respiration rate and the temperature are also noted in the cuts.

The markings and estimates of the time differences are very nearly correct. Cardiograms, especially high-sweep ones, often present confusion in regard to the proper localization of the point which should mark the beginning of systole. At casual view one would naturally locate the point at the bottom of the curve preceding and sweeping up into the main ascending line; but after much attention to this matter it is concluded that in the tracing with a deep depression preceding the main up-stroke the true basal point is situated on the latter near the bottom, and where it may generally be discerned by a slight crook in the line. The line *B*, in Fig. 80, Fig. 82, and Fig. 83, intended to show the beginning of systole, is placed in accordance with this view. To place it through the bottom of the curve would make the time differences in these cases about $\frac{1}{35}$ of a second greater. In Fig. 81 the basal point is more plainly marked, and for the reason that the lever not then on the swing from inertia had become nearly poised, describing a line approaching the horizontal, just before it received the systolic impetus.

The Showings and Interpretations of the Records.

The *form* of O'Leary's pulse-tracing here shown is, in some respects, peculiar, and not such as one, at first thought, would expect to see from a man of such training and endurance. That the form is inherent in his pulse, and was not artificially produced, is proved by testing the instrument as prepared for the observation on K.'s pulse and obtaining his known characteristic tracing with high second wave. The form is not incident to the exercise, because his pulse taken before presented the same type, and tracing Fig. 83, taken forty hours after, does not depart from it. This form, as seen, is characterized by a fairly high first wave, a low second wave, and a prominent third or aortic wave rising from a low aortic notch. As usually described, the form is *dicrotic*. The tracing is interpreted as indicating good heart power and low mean arterial tension. The pressure index at 16 and 17 degrees shows, also, the cardiac efficiency. Again, the ample cardiograms under the conditions of a thick and compact chest-wall testify to the integrity and vigor of the heart.

The pulsations are remarkable for their *regularity of rhythm*. As a general rule, pulses accounted regular, by ordinary tests, undergo changes in rhythm which become plainly manifest by measurement. But in each run here shown the rhythmic changes are scarcely appreciable by measurement. The duration of one pulsation is closely that of another, and the relation between ventricular systole and diastole is closely maintained.

One is struck by the small change produced in O'Leary's pulse by the prolonged and arduous exercise. The changes are, a small increase of rate, of amplitude, and, perhaps, of tension; but these are no more than would occur to any healthy man under ordinary daily conditions.

And as regards the *time differences*, these result from two factors, namely: (*a*) the interval between the contraction of the ventricle and discharge into the aorta, *the presphygmic interval*, as it is termed; and (*b*) the time occupied by the pulse-wave in travelling from the aorta orifice to the radial artery at the wrist. Previous investigations have shown that these factors are variable in value, but the laws governing the changes remain to be ascertained. Under normal conditions, it has been found that the more usual range in adult men is between the sixth and seventh of a second. During the walk O'Leary's time differences between the heart and radial ranged from a sixth to an eighth of a second. The why of this increased range in the direction of shorter time it would be interesting to know. In Fig. 83, taken after the walk, the time differences measure close to the usual two thirteenths of a second.

In the light of these records do we not have a solution of O'Leary's remarkable endurance, and see the reason of his renowned success over all competitors of the late international walking match? His circulation under exercise suffers no strain, no disturbance, but keeps on in the even tenor of its flow and movements. Doubtless the low arterial tension normal to him is an advantage; for the heart thus acting under a minimum of resistance is always equal to its work, never tires, nor is goaded to undue exertion. It would seem that with this efficient well-balanced, and easy-going circulation, and the little disturbance it betrays under such tests, that, but for the want of sleep, he might walk on indefinitely, and his heart evince no signs of failure.

The Pulsations of the Fontanel: Their Form and Mechanism, and Relation to the Pulsations of the Heart and Arteries, and the Movements of Respiration.

The observations from which are derived the facts of the present contribution were made on a healthy male infant, eleven months old, with an open and freely pulsating anterior fontanel. They were made while the child was asleep on three separate evenings. From the glasses bearing the tracings two were selected for reproduction. These with indicatory markings and statements, are represented in the accompanying cuts. Those not published show the same characters and essentially the same relations. The instrument was thoroughly tested, the observations conducted with every care, and results proved and counterproved.

As the time differences vary in value, pains were taken to measure them from several pulsations of a run separately and then compute the average. The results can be read in the illustrations.

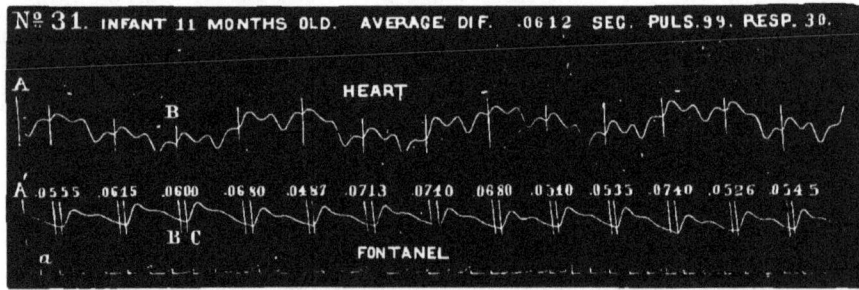

Fig. 84

The range of variation, especially in Fig. 84, is unusually great, which is attributed to misplacement of some of the cardiac basal points by the action of the respiration. However, the measurements being made from the pulsations as they run, the average must efface or render immaterial such errors.

Viewing the representations, first the cardiograms are objects of interest, for good heart tracings of so young a child are by no means common. The entire line undulates with the respiration, rising in inspiration and falling in expiration. A noticeable fact is the relation of systole to diastole, which events are seen to average nearly an equal duration; while in adults the ratio of systole to diastole has been ascertained to be nearly as two is to three.

The tracings of the fontanel next elicit attention. The *form* in general features resembles that of the arterial pulse, although distinctions are apparent which we will not now stop to describe. The *rhythm*, like that of the arterial, is in unison with the heart's pulsation. The *amplitude* is increased at expiration and decreased at inspiration.

But the highest interest attaches to *the chronometric relations and mechanism of the fontanel pulsation.* It is seen that the time difference between the systole of the ventricle and the rise of the fontanel (Fig. 84) averages .0612, or a particle less than one sixteenth of a second. Like measurements from two other well-traced glasses average .0587 and .0562 second respectively; and three other pairs measure each .0555. So the average of all would be .0571 second. No results have been obtained which would make the average time longer than that of Fig. 84. In order to appreciate the brevity of this interval, it is necessary to consider first the distance of the fontanel from the heart by the way of the arteries. It is well known that the course from the heart does not deviate greatly from a direct one (though the course on the right side is straighter than on the left) until the internal carotid is reached. This artery, though usually straight in its cervical portion, is described by Wilson as "remarkable for the number of angular curves which it forms; one or two of these flexures are sometimes seen in the cervical portion of the vessel, near the base of the skull; and by the side of the sella Turcica it resembles the italic letter *s*, placed horizontally." It is safe to assume that the distance by the vertebral arteries to the base of the skull is about equal to that by the carotids. Thus the intercranial arteries, from their source to their distribution in the membranes and periphery, pursue a very devious course, and the distance of the fontanel from the internal carotid foramen can only be approximated by adding increments to the direct line. We have seen that the average time difference of the heart and fontanel by our published showings is .0612 second, and by all our observations .0571 second. This view, then, presents the phenomenon of a long arterial distance between the heart and fontanel, and a remarkably short interval between the pulsations. But this unlooked-for fact is next brought out the stronger by comparison with other distances and time differences coming under observation. Fig. 85 was traced immediately after Fig. 84, and shows the

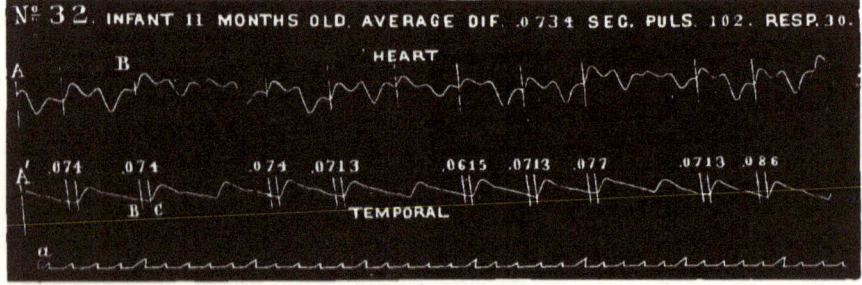

FIG. 85.

interval between the beat of the heart and the pulse of the temporal artery in front of the auditory meatus. The time between the pulsations legibly traced has been computed, and all expressed on the glass and reproduced in the cut. The average time is .0734 = $\frac{1}{13.6}$ second. By these showings the pulse of the fontanel *precedes* the pulse of the temporal artery at the point designated by .0122 or $\frac{1}{87}$ of a second. This is a very nice distinction and, if true, a very remarkable fact. Is it true? The evidences in favor of the precedence of the fontanel over the temporal pulsation are: First, the fact of the capability of the instrument to differentiate so finely. This is unquestionable; it has been sufficiently attested in previous publications, and proofs of the same are constantly appearing in experimentations. Second, the original glasses and their reproduction here presented, in which the measurements were made with the greatest care, the cardiac basal points in both cardiograms being marked at the bottom of the turn into the upward sweep, and the fontanel and temporal basal points at the first beginning of the ascent. Third, another set of glasses, procured on a subsequent evening, which contain beautiful tracings of the heart-fontanel and the heart-temporal pulsations, the cardiograms in even line instead of following, as in the first, the respiratory undulations. These were treated as the first, except that the basal points of both cardiograms were marked at right bearing of the turn into the up-stroke; the effect of which, of course, would be to shorten that much the time differences as computed, but on both glasses alike. This record makes the average time difference of eleven heart-fontanel pulsations .0562 of a second, and the average time difference of thirteen heart-temporal pulsations .0654 of a second. Antecedence of the fontanel to the temporal beat .0092 or near $\frac{1}{108}$ of a second. Fourth, six glasses on which are traced simultaneously the fontanel and temporal pulsations, aggregating fourty-four eligible pairs. The results are too fine for reproduction, but close inspection determines fourteen pulsations occurring without distinguishable difference in time, and thirty-one occurring with antecedence of the fontanel, the latter ranging from a mere iota to the measurement of about a seventy-fifth of a second; but in no instance determines precedence of the temporal pulses. Fifth, the *experimentum crucis* obtained by reversing the order of the tracings. The order is designated *direct* and *reverse;* the former is where the heart or proximal body writes by the upper lever, and the distal body writes by the lower; the latter is where the heart or proximal body writes by the lower lever, and the distal body by the upper. Evidently any change of results, more than the usual range of variation in the same order, obtained by reversing the pulse-bases proves the existence of fallacy, while the production of the same results both in direct and reverse order proves the fidelity of the showings. On three of the above-named glasses the tracings are by reversal with about an equal number of pulsations on either side of the reversal lines; the results of one order agree with those of the other. These glasses have been examined by other competent observers, who have verified the showings as stated.

These must be admitted as very strong evidences in favor of the priority of the fontanel to the temporal pulse ; but against such priority as an established fact, may be advanced, first, admitted liability to fallacious readings and estimates of the showings to an extent at the farthest that would nullify the stated average asynchronism. This liability to error is admitted on the frequent indefiniteness of the basal points, whereby their correct localization within a narrow range becomes uncertain. The difference in question comes within the range, and to that extent its establishment is weakened ; second, the significance and extraordinary nature of the phenomenon in question which would preclude its acceptance on other than incontestable proof.

So, notwithstanding the strong grounds in this case for concluding that the fontanel may, and usually does, precede the temporal pulse in front of the meatus, yet it may not be announced as a fact without further observations. But the following proposition is proved, namely, *that the fontanel pulsates as early as the temporal artery in front of the meatus.*

By the course of the arteries the temporal at point of observation is evidently much nearer the heart than the fontanel, how much nearer it is difficult to closely approximate. The external meatus and internal carotid foramen are nearly on a level ; so if the arteries were equally direct in their course, the temporal in front of the ear and the internal carotid at its emergence from the carotid canal would be about equi-distant from the heart. The deviations of the internal carotid below this point may be counted as equalling at least those of the external carotid and temporal below the height of the meatus. The same comparison is presumed to hold as regards the temporal and vertebral distances. Then, we may say, the heart-fontanel distance is greater than the heart-temporal distance by so much as the entire arterial line running from the internal carotid foramen to the site of the anterior fontanel. The difference would be considerable even if the intercranial arteries ran a course as direct as is usual with other arteries, but the true excess becomes manifest only when the line of their tortuosities is considered. An estimate of four inches would be entirely within the true limits.

If the fontanel pulsation began as late in proportion to distance from the heart as does the average arterial, the first would begin about a sixtieth of a second later than the temporal; the estimate based on a pulse-wave velocity of 240 inches per second and intercranial arterial distance of four inches. If such delay of the fontanel pulse occurred the method would always show it.

In elucidation of the mechanism of the fontanel pulsation the following statement of facts is now in order :

1. *The pulsation of the fontanel begins as early as the pulsation of the temporal artery in front of the meatus.*

2. *The fontanel is some considerable distance by the course of the arteries, say four inches, farther from the heart than the temporal artery in front of the meatus.*

3. *In the arteries on the same line the pulse-wave always arrives later at points more distant from the heart.*

4. *Therefore, the pulsation of the fontanel is not due in its inception to a progressive transmission of the pulse-wave along the intercranial arteries.*

How then is the fontanel pulsation initiated ? Evidently in one of two ways, either the brain rises *en masse* simultaneously with the arrival of the pulse-wave at the intercranial points of the internal carotid and vertebral arteries, or *the intercranial arteries begin their diastole independent and in anticipation of the wave from the heart.* Can we distinguish which of the two theories is correct ? If it be true that the fontanel pulsates *not notably earlier* than the temporal in front of the meatus, then the question cannot be solved by this method, because this relationship would be deemed consistent with either theory. But if, otherwise, it be true that the fontanel pulsates *earlier* than the temporal in front of the meatus, even by so little as one two hundredths ($\frac{1}{200}$) of a second, the question must be considered as solved, for such relationship is alone consistent with the theory that the intercranial arteries dilate before the wave from the cardiac systole can reach them. Hence the great importance of the relationship in question. Happily its determination is within the range of experimental inquiry. The data produced are a contribution, and a sufficient mention of like observations will settle the question beyond dispute. The inquiry, however, would be greatly facilitated by observations on an older subject, in whom the loss of a portion of the cranial bones had left a membranous covering through which the cerebral pulsations could be observed. Such a case as that Dr. Mary Putnam Jacobi experimented on (see *Am. Jour. Med. Sci.*, July, 1878), with a different object from that of this investigation, if subjected to this experimental method, would undoubtedly afford most valuable data, and probably sufficient to settle the point at issue.

Part II.

CLINICAL SECTION.

PRELIMINARY.

Variations of the Presphygmic Interval in Disease, Illustrated by Experiments upon the Schema.

PASSING now to the consideration of cardiac valvular and orificial troubles, we commence with—
Proposition 1.—The presphygmic interval is abnormally shortened in free aortic insufficiency.

François Franck first demonstrated on man that the delay of the arterial pulse on the heart is diminished in aortic insufficiency. The author had independently foreseen the fact, and given its true rationale,* and soon was able to verify its reality by actual observation. The fact needs no further substantiation, and such precipitation of the arterial pulse, notably of the carotid, of course, implies abbreviation or extinction of the presphygmic interval. However, the phenomenon in question is so happily illustrated on the schema that we will not forbear the presentation of two examples of results so obtained.

Fig. 86 shows traces of waves from the ventricle and aortic tube taken with the egress or aortic valve removed, representing free aortic insufficiency. The waves were traced at successively increasing pressures, viz.: 30, 40, 50, and 60 inches.

It is shown that the ventriculo-aortic time difference is inappreciable throughout, the lowest and the highest pressure

* *Cincinnati Lancet and Clinic*, March 29, 1879, p. 224.

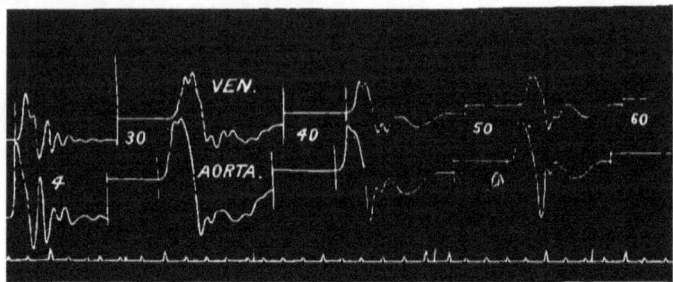

Fig. 86.

giving precisely the same result. It is also shown that the ventricular pressure rises *pari passu* with the arterial.

Fig. 87 represents aortic insufficiency under quick and slow action of the ventricle. It will be noticed that the waves are as nearly synchronous under the one as under the other.

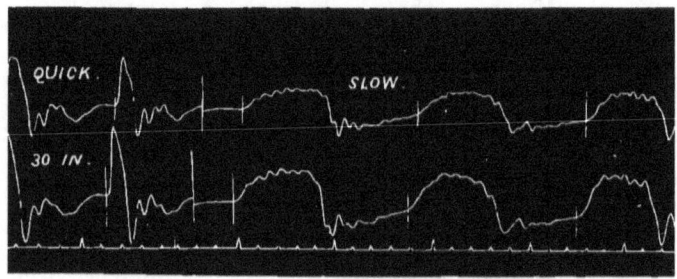

Fig. 87.

These showings are all distinctly different from what has been shown to take place when the valves are intact; and in the light of their testimony we are all the more ready to accept the following explanation of the interesting and valuable diagnostic fact, that the arterial pulse appears distinctly earlier than normal in free aortic insufficiency.

When the aortic valves are permanently open, the blood pressure at the end of diastole is equal in the ventricle and aorta, they constitute parts of one cavity, and, therefore, immediately upon the contraction of the ventricle, the blood pressure in the aorta begins to rise. Whereas, when the aortic valves are intact, the blood pressure at the end of diastole is much lower in the ventricle than in the aorta, and therefore time is required after the beginning of ventricular systole to raise the ventricular pressure above the aortic, which must take place before the arterial pulse can be initiated. In the one case the presphygmic interval is in tangible, in the other it can easily be measured.

In contrast with the preceding is—

Proposition 2.—The presphygmic interval is abnormally lengthened in mitral insufficiency.

We were the first to demonstrate abnormal delay of the arterial pulse in mitral insufficiency. Our published cases show the cardio-carotid interval to be at least double what would otherwise be normal to the individuals. In illustration of this important fact, we will here add one other example:

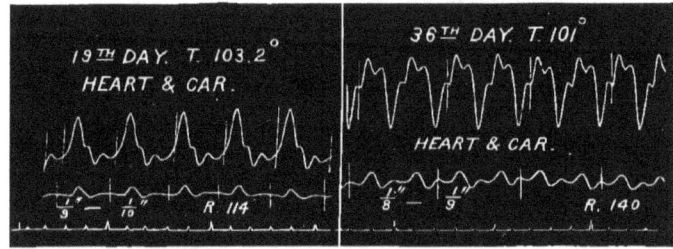

FIG. 88.

Fig. 88 was taken from the patient referred to in Chapter IV., and from whom was taken Fig. 54, showing the carotid-radial traces. It will be remembered he was suffering from typhoid fever complicated with mitral regurgitation. The traces of the first of the figure were taken on the nineteenth day of the fever, with temperature 103.2°, and pulse rate 114. The cardio-carotid interval measures between $\frac{1}{9}$ and $\frac{1}{10}$ second; when, under the conditions, without mitral insufficiency, it could not have measured half as much. The last part of the figure was given on the thirty-sixth day; temperature, 101°; pulse, 140. The cardio-carotid interval here measures between $\frac{1}{4}$ and $\frac{1}{5}$ second, when, irrespective of the valvular lesion, it could not, in any event, have exceeded $\frac{1}{15}$ second. Contrast these intervals with those of the fever cases, Figs. 66 and 70, in which there was no cardiac valvular trouble, and in which the cardio-carotid intervals were not over $\frac{1}{30}$ second.

The schema also is lucid here. If the ingress or mitral valve be removed and a second pouch added to the ventricle, in imitation of the auricle, and then these worked in imitation of the action of the heart, and traces taken, we get a prolonged ventriculo-aortic time difference. Fig. 89 gives results obtained under the conditions named.

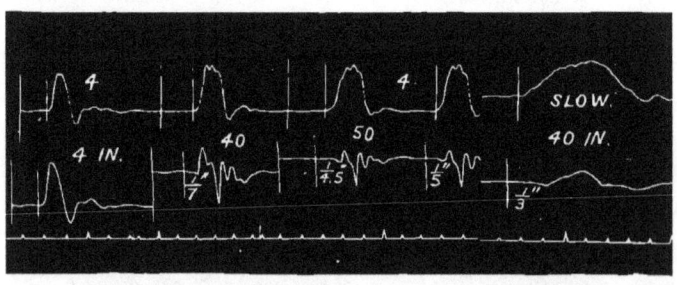

Fig. 89.

These intervals are very long, compared with those of Fig. 68, in which the valves were intact.

From the form of the ventricular traces it might be supposed that the action was slow, and the longer intervals resulted therefrom; but in fact the contractions were quick, and the sloping ascents were in consequence of the free backward escape of the liquid. The traces in the latter part of the figure show the increased lengthening effect of slow ventricular contraction.

Then, with the proofs in its favor, we risk nothing in accepting abnormal delay of the arterial pulse—in other words, elongation of the presphygmic interval, as a certain effect of free mitral insufficiency. And the mechanism of the phenomenon we would explain thus: When ventricular systole begins, there being no mitral barrier, the blood first flows into the relaxed auricle, and is not turned into the aorta until a sufficient head of pressure shall have gathered to force the aortic valves. Time is thus lost between the beginning of ventricular contraction and that of aortic expansion, and the presphygmic interval is accordingly lengthened.

Proposition 3.—The presphygmic interval is lengthened in that variety of aortic obstruction in which the elevation or opening of the valves proves to be difficult independent of the blood pressure.

Examples have been elsewhere contributed of great delay of the carotid pulse arising from this cause, as proved *post mortem*. The mechanism and result can be aptly shown on the schema.

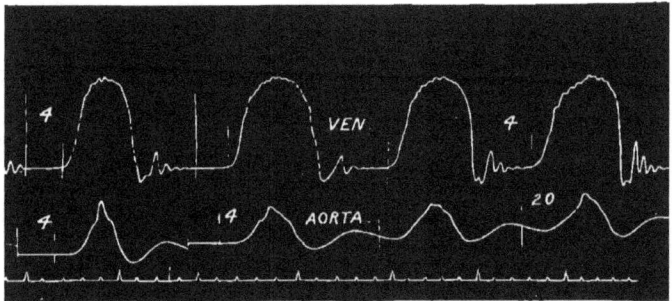

FIG. 90.

Fig. 90 was procured with the egress valve pressed upon by a spring-weight, which permitted it to yield only to a superior pressure. It will be observed that the ventriculo-aortic intervals are very long, and even with the liquid pressures in equilibrium at 4 inches. In the figure, the effect of stenosis is added to that of the heavy valve, as shown in the sloping ascent and rounded and distant summit of the arterial trace. But it is possible to obtain on the schema what sometimes happens in the living, viz.: retardation of the beginning without retardation of the summit, by having the valve close with a species of locking, so that its opening is delayed, but when once forced it rises up freely. Thus heavy aortic valves without stenosis cause delay of the beginning without delay of the summit of the arterial trace; while heavy valves with stenosis cause delay both of the beginning and summit. Manifestly the presphygmic interval is only concerned in delay of the beginning of the arterial wave.

Proposition 4.—Aortic obstruction from pure aortic stenosis does not cause elongation of the presphygmic interval, but only delay of the arterial summit.

It is plain that, with pliable aortic valves, the blood would begin to flow as soon through a small as through a large orifice.

The form of the pulse in aortic stenosis is familiar to all, but Fig. 91, here republished, shows not only the peculiar form, but that the beginning of the pulse was not delayed.

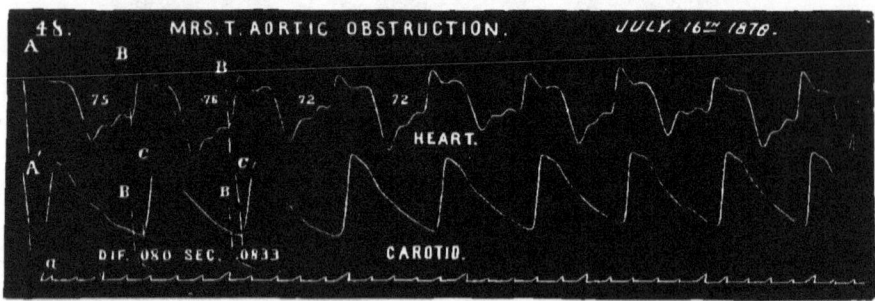

FIG. 91.—EFFECT OF AORTIC STENOSIS.

Schema: In Fig. 92 the traces of the first part were taken at thirty-five inches pressure, with the valve intact and tube free; and the traces of the last part at thirty inches pressure, with valve intact and tube constricted by compression just in front of the valve. It is seen that while the form of the arterial trace after the compression is strictly that of aortic stenosis, the beginning of the waves are not in the least later than before the compression.

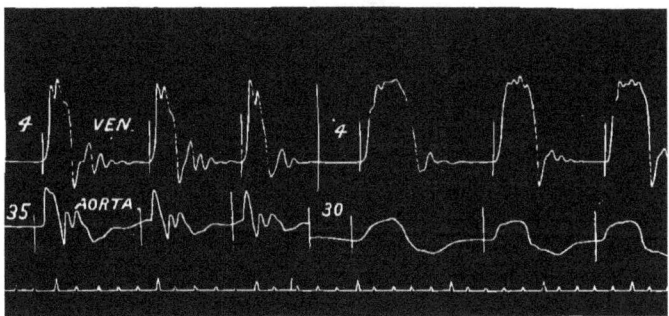

FIG. 92.—EFFECT OF AORTIC STENOSIS SHOWN ON THE SCHEMA.

Therefore the presphygmic interval is not lengthened in aortic stenosis if the aortic valves are pliable.

Proposition 5.—The effect of mitral constriction on the duration of the presphygmic interval remains as a problem to be solved.

No examples have been furnished from any observer of measurement of the cardio-arterial interval in cases of mitral stenosis. Experimental data in relation to this point are derived alone from the schema, and these, though positive, we hesitate to accept until confirmed by observations on man. However, the fact is patent that in no instance has the schema failed, when applied, of reproducing the same effects as observed in man.

Fig. 93 shows the result obtained, the first part under normal conditions, the last part with ingress tube constricted immediately behind the valve. The ventricular trace indicates that under the constriction the pressure within must have fallen in diastole to a point relatively low. But the striking showing is the great delay of the arterial trace, showing about $\frac{1}{4}$ second, whereas the normal shows about $\frac{1}{10}$ second.

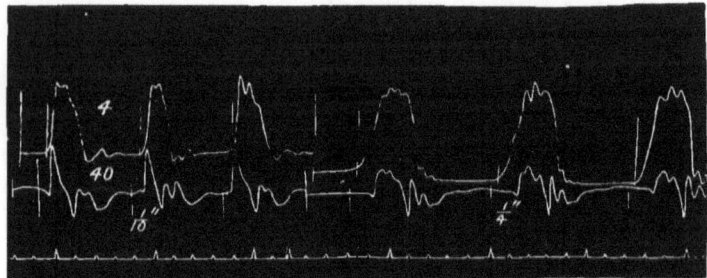

FIG 93.—EFFECT OF MITRAL CONTRACTION SHOWN ON THE SCHEMA.

This result was a surprise; and yet, on maturer reflection, it does not appear inconsistent with the mechanisms involved in mitral contraction. At the end of diastole, the ventricle being quite relaxed and unfilled, and systole starting under these conditions, it would progress longer than usual before the parietes would press sufficiently upon the contents to force them through the aortic valves. With the mitral orifice and valve normal, systole begins upon a distended ventricle, the pressure rises rapidly from the start, and the overflow into the aorta promptly begins.

Therefore we conceive that the problem *will* be demonstrated in favor of the ability of mitral constriction to produce exaggerated delay of the pulse, which implies elongation of the presphygmic interval.

The following are prominent among the facts determined by this last research:

1. The duration of the presphygmic interval is *increased* in slow ventricular contraction; infrequent pulsations; relatively high arterial pressure; heavy aortic valves; mitral insufficiency; and probably mitral contraction.

2. The duration of the presphygmic interval is *diminished* in quick ventricular contraction; frequent pulsations; relatively low arterial pressure; and aortic insufficiency.

CHAPTER VII.

DIMINUTION OF RETARDATION OF THE PULSE IN AORTIC INSUFFICIENCY—THE INFLUENCE OF AORTIC ANEURISM AND AORTIC INSUFFICIENCY, SINGLY AND COMBINED, ON THE RETARDATION OF THE PULSE.

Aortic Insufficiency.

DR. HENDERSON,[*] in 1832, first emitted the idea that the interval between the heart's impulse and the arterial pulse is prolonged in aortic insufficiency. Since him others, and among them the principal authorities on diseases of the heart, have accepted the view. Flint says: "That it characterizes certain cases in which the regurgitation is excessive, is not to be denied."[†] Walshe writes: "This retardation may, with care, be detected in many, but unquestionably not in all, cases of that disease. Possibly, where no morbid retardation can be discovered, the failure may depend not on its absence, but on its being carried to such extremes that the arterial pulse produced by one cardiac systole is nearly synchronous with the next.[‡]" As late as 1877, M. Tripier, in a publication, § advocated the reality of this exaggerated delay of the pulse in aortic insufficiency.

The observation appeared incontestable. The hand perceived the shock of the heart, and the finger the radial pulse, the interval between the events being noted as much longer

[*] *Edinburgh Medical and Surgical Journal*, vol. xxvii., 1832.
[†] Flint's "Diseases of the Heart," 1859, p. 141.
[‡] "Practical Treatise on Diseases of the Heart and Great Vessels," American edition, 1862, p. 72.
§ *Revue Mensuelle*, p. 19.

than in health. And the explanation of the accepted phenomenon came with show of reason, through the reverse arterial current and lowered arterial blood pressure (classical) of free aortic regurgitation. The view of an abnormal delay of the pulse, thus supported by observation, reason, and authority, seemed an established fact in the clinical history of this lesion.

Nevertheless the idea is wholly erroneous, and the pulse, so far from being unduly retarded on the systole of the ventricle, is really greatly precipitated on that event in large aortic insufficiency. Correction of the prevailing error, and demonstration of the true chronometric relationship between the heart and pulse, is due to the graphic method. Traces of the heart and an artery taken simultaneously, show neatly the beginning of cardiac systole and the beginning of the arterial pulse, and the space separating these beginnings marks definitely the interval between the events. Thus the normal interval between the heart and different arteries being ascertained, the modification of the interval by disease is readily noted.

In this manner François Franck experimented on patients affected with aortic insufficiency, and first presented his results to the Société de Biologie in March, 1878. He formulated thus the conclusion of his researches: "*In large, pure aortic insufficiency the retardation of the pulse on the beginning of the systole of the heart is very notably diminished.*" Co-temporaneously with Franck, the author also was studying, independently, by means of the simultaneous graphic method, the influence of different forms of valvular disease on retardation of the pulse, and had demonstrated that the pulse is abnormally delayed in mitral insufficiency, and reflecting as to whether this delay might be contravened by any concomitant condition, arrived at a conclusion which was afterwards published in the following words: " Nevertheless, the phenomenon, though constant in pure mitral incompetency, will probably be found wanting in cases of this lesion complicated with an open state of the aortic valves; for in the latter condition the base of the arterial column rests against the sides of the ventricle, instead of against the aortic valves, and is advanced with the first movement of ventricular contraction, thus insuring a short interval between cardiac systole and arterial expansion. . . . The idea that aortic insufficiency produces delay of the pulse is certainly erroneous." *

In March, 1880, opportunity presented for tracing a case of undoubted aortic insufficiency. The result is shown in Fig. 94.† The cardio-carotid interval measured only $\frac{1}{35}$ second, and the cardio-radial $\frac{1}{13}$ second.

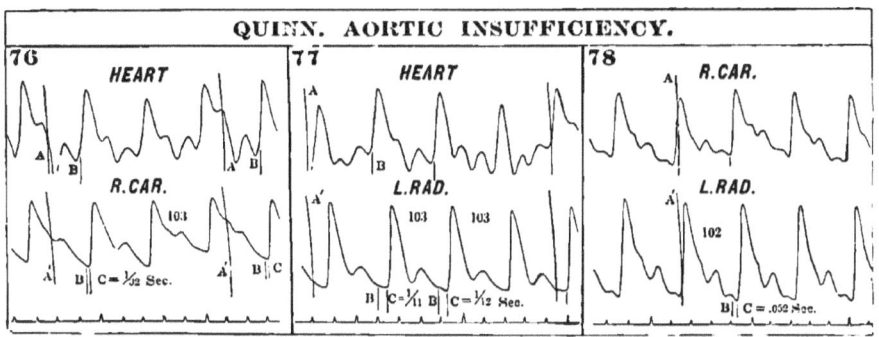

FIG. 94.—SIMULTANEOUS TRACES IN A CASE OF LARGE AORTIC INSUF-
FICIENCY, SHOWING ABNORMAL SHORTENING OF THE CARDIO-
CAROTID AND CARDIO-RADIAL INTERVALS.

Still referring to his own researches, the author has proved diminution of delay of the pulse in other cases of aortic insufficiency, and has been able, also, to reproduce the same result on the schema. By way of illustration, simultaneous tracings from the schema are here presented. Fig. 95 shows traces from the ventricle and aorta, with valves intact. The impulses were given to the ventricle at successively increasing arterial pressures. It will be noticed that after the first pair of waves, at which the pressure was in equilibrium and the waves are synchronous, the interval between the ven-

* *Cincinnati Lancet and Clinic*, March 22, 1879.
† *Boston Medical and Surgical Journal*, September 30, 1880.

tricular impulse and arterial wave increases with the arterial pressure. Thus, with the pressure at 30 inches (water

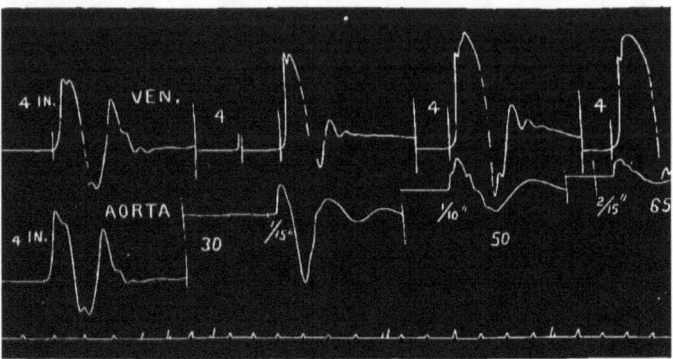

Fig. 95.

manometer), the interval is $\frac{1}{5}$ second, at 50 inches $\frac{1}{6}$ second, and at 65 inches $\frac{2}{15}$ second. The traces represent normal action and lengthening of the cardio-arterial interval under augmentation of arterial pressure.

Fig. 96 shows the result of a repetition of the same experiment, only with the important difference that the traces were taken with the aortic valve removed, in representation of large aortic insufficiency. Here the ventricular and aortic

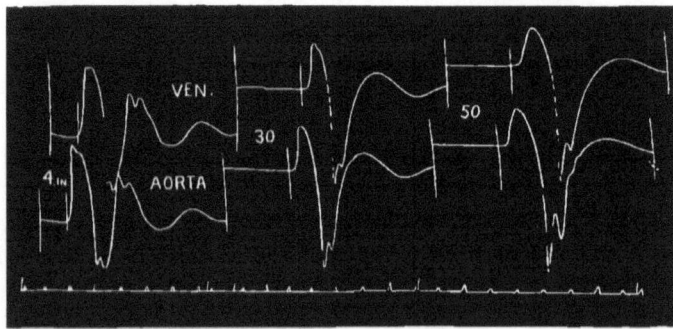

Fig. 96.

waves are synchronous throughout, and, besides, show a perfect parallelism of form.

Franck also experimented on animals, producing in them artificial insufficiency of the aortic valves, and proved the same diminution of the pulse-retardation observed as in man.

Thus experiments on man, on animals, and on the schema all concur in proving diminution of delay of the pulse in aortic insufficiency. The testimony is ample and complete, and establishes the fact beyond question. The acquisition is one of the triumphs of the graphic method, and affords a striking illustration of its power to redeem from error which otherwise had been perpetuated, and reveal the truth which otherwise had not been known.

The Mechanism of the Phenomenon.

The next inquiry relates to the mechanism of the phenomenon. The conditions concerned that influence the measure of the delay of the pulse are (1) states of arterial blood pressure ; (2) states of the arterial coats as to elasticity or stiffness ; (3) modes of ventricular contraction ; and (4) relative states of ventricular and aortic blood pressure.

1. It has been found that, other things being equal, the rate of propagation of the pulse-wave tends to increase with increase, and decrease with decrease, of arterial blood-pressure. Recent investigations, however, show that such modification from such cause is small at best, and frequently fails to manifest. In aortic regurgitation the mean arterial pressure is usually diminished, yet not unfrequently, when the lesion is well tolerated, the pressure maintains, or even rises above, its normal level. This cause, then, of modification of pulse transmission, would operate either against, or inadequately for, the production of the greatly abbreviated interval.

2. No fact in this relation is better established than that the velocity of the pulse wave is proportional inversely to the elastic extensibility of the arterial walls. In aortic insufficiency great expansion of the arterial walls is a notable

phenomenon. This is due to the enlarged and hypertrophied ventricle sending into the arterial system at each systole a large volume of blood, which distends the vessels, and in diastole escaping backwards permits corresponding retreat of the walls. The walls themselves may not be more relaxed or extensible than ordinary, and the fact is the coats are often pervaded with atheromatous material, rendering them less yielding than in normal condition. However, in the absence of indurative changes, the parietes under the strain become dilated and thinned, and more yielding, unless fortified by developing hypertrophy of the muscular layers. Then no constant stiffening of the arterial walls obtains in aortic insufficiency on which can be predicated the very early appearance of the pulse in this lesion. But even if the walls were of brass, this cause of quickened transmission could not of itself account for the great precipitation of the carotid pulse, inasmuch as the reduction considerably exceeds the entire transmission interval between the heart and carotid point.

3. The mode of ventricular contraction, as quick or slow, exerts a marked influence on the amount of delay of the pulse. The author's recent experiments have shown this, but they have also shown that the modification is confined alone to the presphygmic interval of the systole of the ventricle. Contrary to the conclusion of some other observers, the author's results demonstrate that the velocity of the pulse wave is not in the least modified by the quickness or slowness of ventricular contraction. The ventricular presphygmic interval is notably shortened by a quick and lengthened by a slow contraction of the ventricle. In aortic insufficiency, although the coexisting hypertrophied left ventricle contracts with corresponding energy, there is no evidence that its initial systole is quicker than in normal action. Then mode of systole, notwithstanding its modifying potency, cannot certainly be invoked in explanation of the phenomenon in question.

4. It is the excess of blood pressure in the aorta over that in the ventricle at the beginning of systole that measures

with certainty the interval between the beginning of ventricular contraction and that of the aortic pulse. When aortic valves are intact this interval lengthens with increase and shortens with decrease of this excess; and if the pressures should be in equilibrium the two events will begin simultaneously. (See Fig. 95.) But when the aortic valves are permanently open, the pressures in the ventricle and aorta are always in equilibrium at the beginning of systole (the two cavities being in one), whether the mean pressure be low or high. In consequence of this oneness or equilibrium of pressure the heart's impulse and aortic pulse will be synchronous. (See Fig. 96.)

The transmission interval between the aortic orifice and point of observation of the carotid pulse is comparatively short, not averaging more than half the duration of the cardio-aortic or presphygmic interval. Thus, by measurements, the first averages .026 second, and the last .054 second, with pulse rate at 75. In permanently patulous aortic valves the larger presphygmic interval is practically obliterated, while only the smaller cardio-carotid transmission interval remains as the delay of the carotid pulse on the heart.

The above is the true explanation, and to the author is due the credit of its first recognition and announcement. Franck at first did not grasp the mechanism of the phenomenon he had proved, but vainly endeavored to account for it by the theory of accelerated transmission of the pulse wave. His later deliverances, however, on this point, are in perfect accord with the true rationale here set forth.*

It is easy to explain the fallacy of an exaggerated delay of the pulse on the systole of the ventricle in aortic insufficiency. The enlarged ventricle suddenly filling from both the aorta by reverse and the auricle by direct flow, communicates a shock so marked as to be mistaken for systole. This impulse occurring in the first part of diastole, and preceding the arterial pulse at such distance, gives the impres-

*Compt. Rend. de la Société de Biologie, 27 Janvier, 1883.

sion of enormous delay of the pulse. In Fig. 96 the length and steepness of the diastolic ascents show how easily the diastolic impulse would be taken for the systolic beat. The fallacy arose from an error of observation, which the graphic method was needed to correct.

The Diagnostic Value of the Phenomenon.

Excessive diminution of delay of the arterial pulse, notably the carotid, is a sign of the highest importance in large aortic insufficiency. No other condition, or combination of conditions, except one to be considered, is capable of producing such marked precipitation of the pulse on the heart. The exception consists in the combination, found only in fever, of quick systole and tendency to equilibrium of ventricular and aortic blood-pressures, these conditions also invariably associated with frequent pulsations. The febrile condition, then, with quick cardiac systole, frequent pulse, and low arterial blood-pressure, is capable of reducing the cardio-carotid interval, the same as aortic insufficiency, to the value of the transmission time. These factors, when present, are well declared, and in their absence the graphic sign is pathognomonic of the lesion in question. In positive value it outranks diastolic basic murmur; which, as is well known, may originate in the aorta without regurgitation, or again with only slight regurgitation, which takes place in the first part of diastole, whilst the altered valves are falling into position of closure. Besides, there is difficulty sometimes in distinguishing between aortic regurgitant and mitral direct murmur of the first part of diastole.

As to default of this sign, in large aortic insufficiency it never fails, except perchance in the presence of an extensible aneurism of the first part of the aorta. The fact that an aneurism with yielding walls produces delay of the distal pulse has been well established in these recent years. Then, in a case of large aortic insufficiency, complicated with a yielding aneurism of the ascending aorta, the precipitating effect of the valvular lesion would be more or less counterbalanced by the retarding effect of the aneurismal pouch, and the carotid pulse would observe, or approach, the normal amount of delay. If aortic aneurism be eliminated, the presence of abnormal precipitation of the carotid pulse is conclusive, in any case, of an open state of aortic valves.

From this declaration it is plain that the author does not admit, with François Franck, default of the sign in the presence of concomitant cardiac valvular lesions. Coexisting aortic stenosis would give the characteristic sloping ascent, but the beginning of the pulse would be in no wise delayed. This has been demonstrated on man and the schema. Mitral regurgitation coexisting, auricle, ventricle and aorta would constitute one cavity, with blood pressure in equilibrium at the end of diastole, and the blood would be as promptly sent forward into the aorta as in pure aortic insufficiency. Nor could mitral contraction, if present with aortic insufficiency, cause any delay of passage of blood from the ventricle, as the pressure in the latter at the end of diastole would always be equal to the aortic pressure. Hence, whatever the cardiac complication, there is no failure of abnormal precipitation of the beginning of the pulse as a sign of permanent aortic insufficiency.

But the fact must not be lost sight of that this sign, so positive and constant in large aortic insufficiency, will fail to manifest in the form of incomplete lesion, in which the valves permit of regurgitation in the first part, but effectually close in the last part, of diastole. In this state of things, when systole begins, the valvular barrier and excess of aortic blood pressure being present, time is lost in overcoming the resistance, and abnormal precipitation of the pulse fails to occur. However, default of the sign in incomplete insufficiency is more than compensated in diagnostic import, in that the absence of abnormal pulse precipitation in a case of aortic regurgitation, certainly diagnosticated by the ordinary physical signs, would indicate a partial and not complete insufficiency of the valves; and, aortic aneurism excluded, would be conclusive of this distinction.

In *résumé :*

1. Abnormal diminution of the retardation, or, in other words, abnormal precipitation of the arterial pulse, notably the carotid, on the systole of the ventricle in large aortic insufficiency, is a fact positively established, the phenomenon depending purely upon extinction of the normal ventriculo-aortic, or presphymic interval. Hence the phenomenon becomes an important diagnostic sign of this lesion.

2. The presence of the sign is positive evidence of the existence of the lesion, provided only there is no quick febrile movement in the case.

3. Default of the sign does not occur in the presence of concomitant cardiac lesions, but occurs only in the presence of a yielding aortic aneurism. Hence,

4. Absence of the sign is positive evidence of absence of the lesion, provided only there is no aortic aneurism in the case.

5. If the diagnosis of aortic regurgitation is otherwise certain, absence of the sign, aortic aneurism eliminated, is positive evidence of the incomplete nature of the insufficiency.

The Influence of Aortic Aneurism and Aortic Insufficiency, Singly and Combined, on the Retardation of the Pulse.

" L'anévrisme de la portion initiale de l'aorte coexistant avec une insuffisance aortique large, on n'observe pas l'exagération générale du retard du pouls qui est caractéristique de l'anévrisme siégeant à ce niveau ; l'influence retardatrice de l'anévrisme est contre-balancée, par l'influence inverse de l'insuffisance aortique ; la résultante de ces deux effets opposés qui se combinent, est la conservation de la valeur normale ou presque normale du retard du pouls."*

In the above paragraph François Franck states in effect three distinct propositions:

* *Journal de l'Anat. et de la Physiol.*, t. xv. (Mars-Avril, 1879).

(1) Aneurism of the ascending portion of the aorta increases the retardation of the pulse above the normal value.

(2) Large insufficiency of the aortic valves diminishes the retardation of the pulse below the normal value.

(3) The coexistence of aortic aneurism with large aortic insufficiency causes the retardation of the pulse to preserve the normal, or nearly normal, value.

The truth of the first proposition has been sufficiently attested by François Franck's and the author's researches. That aortic aneurism produces abnormal delay of the pulse is no longer a question, but remains as one of the great facts of clinical sphygmography.

In regard to the second proposition, François Franck first insisted on theoretical grounds that the pulse would appear earlier than normal in the condition of large aortic insufficiency, and afterwards was able to produce clinical cases in proof of the correctness of the view. The author also, before he had known of M. Franck's researches, had arrived by *a priori* reasoning at this same conclusion, and, in substantiation, places on record the following case :—

Cincinnati Hospital. Service of Dr. C. G. Comegys. Quinn, age thirty-two years. Diagnosis, aortic insufficiency, with hypertrophy of left ventricle. Graphic records taken March 27, 1880, as shown in Fig. 94. No. 76 shows simultaneous tracings of the heart and right carotid, and No. 77 of the heart and left radial; the corresponding time in each also is shown. By very careful measurement the delay of the carotid pulse on the heart is placed at one thirty-second of a second. The normal delay with pulse rate at 103, as in this instance, would be about twice this interval. The cardio-radial interval is shown to measure one eleventh to one twelfth of a second. The normal interval would be not less than one eighth of a second.

This was a plain case of free aortic regurgitation, and the graphic records can only be regarded as demonstrating a greatly abbreviated cardio-arterial interval as the direct effect of the aortic insufficiency.

What is the explanation of this interesting fact? In large aortic insufficiency the base of the arterial column rests against the sides of the ventricle instead of against the aortic valves, and is advanced, causing rise of the pulse with the first movement of ventricular contraction. The result is due to shortening or near extinction of the presphygmic interval.

The velocity of the pulse-wave in this case was rapid indeed, as shown in No. 78, where the time difference between the carotid and radial pulses measures about one nineteenth of a second. But the greatest possible reduction of the cardio-carotid transit interval would signally fail to account for the great reduction here instanced of the cardio-carotid interval. Therefore the reduction must have been due to marked abbreviation of the presphygmic interval.

In practical value this fact, it is believed, stands first among the list of clinical facts brought to light by the graphic method. It stands as the exponent of large aortic insufficiency. This condition and no other will produce abnormal precipitation of the pulse on the heart.

In this connection, however, it is important to observe that the phenomenon under consideration will only be manifest in complete (large) insufficiency of the aortic valves—the condition in which they are permanently open,—and the beginning of ventricular systole finds no barrier between the blood in the ventricle and the blood in the aorta. The condition of partial insufficiency, in which the valves permit reflux for a time after the beginning of diastole, yet finally close so that the beginning of systole finds the barrier intact, will not signal the phenomenon.

The third proposition is necessarily deducible from the two preceding. If aortic aneurism causes retardation of the pulse, and aortic insufficiency causes precipitation of the pulse, then the conjoint action of the two conditions would counteract each other, and cause the succession of the pulse on the heart to approximate more or less nearly the normal time.

DIMINUTION OF RETARDATION OF THE PULSE. 171

The following case is given in exemplication of this interesting and important fact :

Charles Gardner, aged fifty-six years, came under notice in November, 1878, presenting the following symptoms and physical signs: General weakness, dyspnœa on exertion, inability to lie down, pains in chest, back of head, left side of neck, and left shoulder; pulse to fingers strong, full, frequent, and regular; heart's impulse increased; area of cardiac dulness increased; pronounced double murmur heard over front of chest and in the back; the systolic accentuated at the right base, and conducted towards right clavicle; the diastolic also ac-

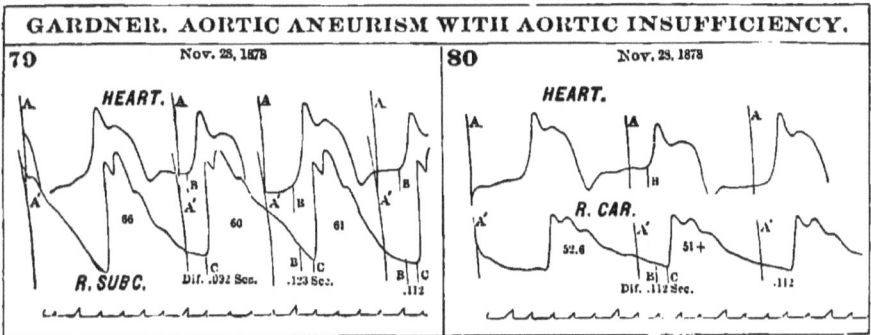

FIG. 97.

centuated at right base, but propagated downwards over body of left ventricle ; murmur but faintly heard at apex ; at the site of the right subclavian pulse, above the clavicle, a pulsating tumor the size of a pigeon's egg ; also visible, but less prominent, pulsation of left subclavian. Graphic observations with the compound sphygmograph were taken November 28th. Samples of the tracings are here faithfully reproduced. In Fig. 97, No. 79 shows the heart and right subclavian tumor; No. 80, the heart and right carotid ; in Fig. 98, No. 81, the right subclavian and left

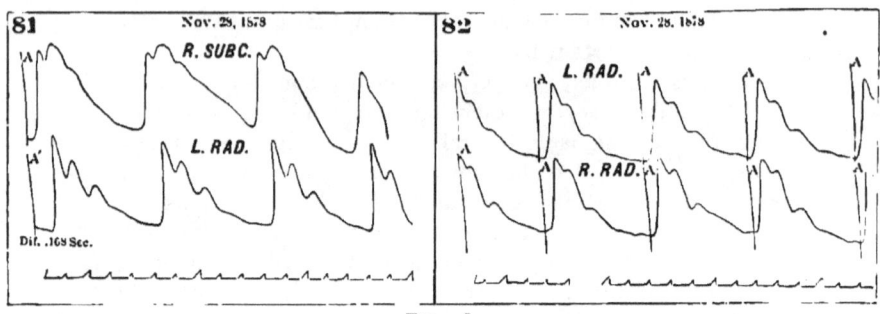

FIG. 98.

radial; and No. 82, the two radials,—each pair taken simultaneously, with accompanying time in fifths of seconds.

In December, 1879, the case was noted again. There had been progressive decline, the pains had become more severe and persistent. The action of the heart and arteries had greatly diminished, and the pulsations become more frequent. The subclavian pulsations had subsided within normal limits. The evidences of enlarged heart, and the murmurs, with the same characters, continued; and in addition feeble undulations were perceptible in the second right interspace near the sternum, and in the first and second left interspaces. Also, the voice at times showed failure, and the power of deglutition was not always perfect. December 20th, tracings were obtained as shown in Fig. 99, Nos. 83 and 84; again, Jan-

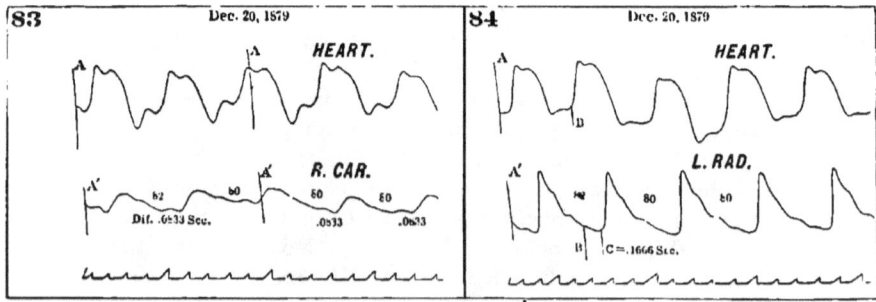

FIG. 99.

uary 7, 1880, as shown in Fig. 100, Nos. 85 and 86, and January 17th, as shown in No. 87.

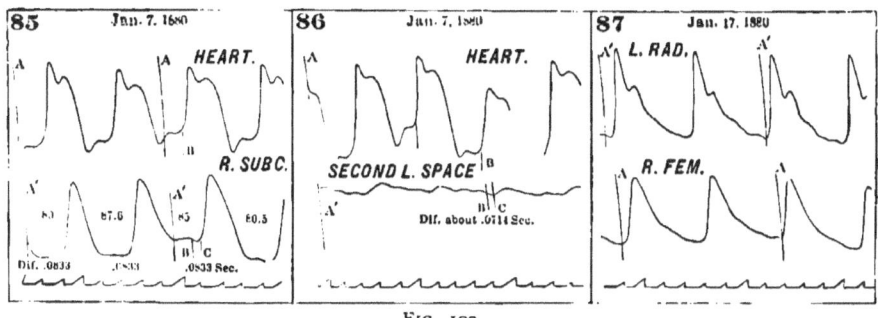

FIG. 100.

The patient continued to decline, and died July 13th, 1880, right hemiplegia setting in a day or two previous to his death.

At the post-mortem were present Drs. Isham, Freeman, and Barrows. The following are the notes: Pericardium contained several ounces of fluid. Left ventricle hypertrophied, walls fully one inch thick. Mitral valves, with the exception of slight thickening towards the base, healthy; evidently fully competent during life. Aortic orifice enlarged; aortic valves thickened, corrugated, and calcified throughout; evidently wholly incompetent during life. Ascending aorta greatly dilated, atheromatous, and studded with calcareous matter. Transverse portion of arch the seat of a large aneurism. The sac projected from the anterior superior portion of the wall, and was attached to the upper portion of the sternum, the left clavicle, and first left rib, and pressed upon the trachea and œsophagus. It was full three inches in interior diameter; its walls were soft and flabby, about the thickness of normal left auricular parietes. The surface of the sternum, against which the tumor pressed, was eroded, and about half its thickness removed; also, there was erosion of the sternal end of the left clavicle and first left rib. A mass of laminated fibrine the size of a walnut occupied the distal portion of the sac and projected into the descending aorta. The latter was also much dilated. The innominate and first part of the right subclavian were dilated to about twice the normal diameters. The right and left common carotids and left subclavian were about normal size. Both auricles, the right ventricle, and right-side valves were normal.

This case was complete, and a typical one of combined aortic aneurism and aortic insufficiency. The sphygmographic records were also complete. The latter we examine with interest, especially at present, with reference to the succession of the arterial pulses on the systole of the ventricle. Evidently aneurism, as in the present case, of the transverse portion of the arch, involving the origins of the three great arteries, namely, innominate, left common carotid, and left subclavian, entails the same effect on the succession of the pulses as aneurism of the ascending portion only; that is, the pulses are all abnormally delayed, and the symmetrical pulses of either side are equally delayed.

The tracings were taken and the successions measured with the greatest care; so the intervals expressed on the plates are reliable, and must approximate throughout very closely the true ones. The showings are as follows:

November 28, 1878. No. 79. Pulse rate 66–60. Cardio-subclavian interval .109 second (average).
No. 80. Pulse rate 52–51. Cardio-carotid interval .112 second.
December 20, 1879. No. 83. Pulse rate 82–80. Cardio-carotid interval .0833 second.
No. 84. Pulse rate 82–80. Cardio-radial interval .1666 second.
January 7, 1880. No. 85. Pulse rate 87–80. Cardio-subclavian interval .0833 second.

These measurements are sufficiently near the normal. In the variations of the time differences, however, there was noticed a bearing towards the longer time, and the average of all the observations, of which a portion only are shown, would indicate rather longer time than obtains in strictly normal conditions, but not sufficiently prolonged to exceed the healthy range or permit the counting of such delay as a definite indication. Then in this case the coexisting aortic aneurism and aortic insufficiency so counterbalanced each other that the retardation of the pulses on the heart was within the normal range.

According to the size and yielding nature of the aneurism must be the amount of delay impressed upon the pulse. An aneurism may be such in these respects as to cause a delay that would overbalance the precipitation entailed by the aortic insufficiency. On the other hand, the conditions may be reversed, and a large aortic insufficiency entail a precipitation that would overbalance the delay caused by the aortic aneurism. However, when the two conditions coexist, the delay will never be so great as to definitely favor the wrong diagnosis of aortic aneurism alone, nor so small as to definitely favor the wrong diagnosis of aortic insufficiency alone. *If the diagnosis of aortic aneurism be certain, and the pulse, notably the carotid or subclavian, shows no abnormal delay, the conclusion is justified that aortic insufficiency coexists. If the diagnosis of large aortic insufficiency be certain, and the pulse, notably the carotid or subclavian, shows no abnormal precipitation, the conclusion is positive that aortic aneurism coexists.*

So in this case the ordinary signs and symptoms, at the first examination, were sufficient to determine the presence of large aortic insufficiency. Accepting this, the want of abnormal precipitation of the carotid and subclavian pulses enabled the author to arrive at the diagnosis of coexisting aortic aneurism before there were any other indications of this condition.

As of interest relating to the localization of aortic aneurism: No. 82 shows the two radial pulses simultaneous, which proves that the aneurism was seated in the ascending portion anterior to the origin of the innominate, or that it was seated in the transverse portion so as to include the origins of both the innominate and left subclavian (and of course also the left common carotid). No 87 shows the left radial as compared with the right femoral slightly delayed.

This is the more usual normal relationship of these pulses, and proves that the aneurism was anterior to the origin of the left subclavian artery. Evidently, had the aneurism been below the origin of the left subclavian, the femoral would have shown delay as compared with the radial. No. 86 shows that the pulsation in the second left interspace succeeded the cardiac pulsation about .0714 of a second, and therefore aided the diagnosis of aneurism as located.

No. 80 is given to show a very ample and finely delineated radial pulse, and the rate of transmission of the pulse-wave between the subclavian and radial points under the conditions then existing in the case. The time difference measures .108 second, which indicates a rather slow pulse-wave transmission. If this be compared with the carotid-radial interval in the case of Quinn, namely, .052 second, as shown in No. 78, a remarkable difference becomes apparent in the pulse-wave velocity of the two cases. The causes of this diversity are to be found in the dissimilar conditions which pertain to the respective cases.

CHAPTER VIII.

RETARDATION OF THE PULSE IN MITRAL INSUFFICIENCY—ENORMOUS RETARDATION OF THE PULSE FROM MITRAL INSUFFICIENCY, AORTIC ANEURISM, AND HEAVY AORTIC VALVES.

THIS subject, in its inception and literature, seems to belong entirely to the author. For although none would question the importance of a new diagnostic sign of a lesion so common and grave as mitral insufficiency, and whose diagnosis by means of the old signs is not always infallible ; and although he presented to the profession some years ago demonstrations of this phenomenon of mitral regurgitation, which to him were conclusive, the announcement has attracted but little attention ; his observations have neither been confirmed nor disproved by others, and two writers only have noticed them. Dr. A. B. Isham, in presenting the value of the graphic method from the point of view of examinations for life insurance, accepts abnormal delay of the pulse as a valuable sign of mitral regurgitation.* And M. Pélix in his inaugural thesis on retardation of the pulse, quotes largely and approvingly from the author's published reports. † These papers contain all that has been contributed upon the subject by others than the author.

Evidences of the reality of the phenomenon are drawn (1) from a consideration of the mechanism of mitral insuffi-

* *American Journal of the Medical Sciences*, July, 1882, p. 119.
† Paris, 1882.

ciency ; (2) from experiments on the schema ; (3) from experimental observations on man.

Examining the question first from the point of view of the mechanism of mitral insufficiency, we shall find a convincing *a priori* argument in favor of abnormal delay of the pulse as a direct effect of this lesion. Under normal conditions with mitral valve intact, the heart is a perfect organic pump, the left ventricle alternately and duly receiving and sending out blood. When the ventricle contracts, the pressure within becomes positive in relation to the arterial and auricular, which insures the prompt closure of the auriculo-ventricular valve, and immediately thereupon the opening of the aortic valves and escape of blood into the aorta.

When the ventricle relaxes, the pressure becomes negative with respect to the arterial and auricular, and the blood flows freely in from the auricle, but is prevented doing so from the artery by closure of the similunar valves. Auricular systole raises somewhat the diastolic intra-ventricular pressure, but the moment of beginning systole finds a considerable excess of pressure on the side of the artery. The time required to overcome this excess after the beginning of ventricular systole measures the duration of the ventriculo-aortic or presphygmic interval, which under the normal conditions will give a normal value.

Thus all goes smoothly on ; the rhythmic succession of contraction and relaxation, of valvular opening and closure, of inflow and outflow of blood, and all the events of a cardiac cycle in harmony with each other in time and status and movement.

But when the mitral valve is permanently open the harmony is no longer preserved. The ventricle fills in diastole even more promptly and completely than in normal conditions, in consequence of falling into it of blood from the surcharged auricle ; but in systole the blood is first sent backwards through the open valve into the relaxed auricle. This process fills the auricle and pulmonary veins and dams back the blood towards the lungs. The retrograde current, however, is soon arrested by meeting the direct current from the right ventricle, whose systole is synchronous with the left, and whose tricuspid valve is intact. Thus the pressure within the left ventricle and auricle (now one cavity) finally becomes positive with respect to the aortic pressure, and for the remaining part of the systole blood escapes into the aorta. From this it follows that the time and work of each systole is consumed in first driving the blood backwards, and only a part of the systole is effective in driving the blood into the general arterial system. The effect of this condition of things upon the arterial pulse is to postpone the time of its appearance and lessen the volume of its wave; the cardiac systolic portion is shortened relatively to the length of the cardiac systole. If the heart maintain its vigor the pulse may still feel sharp to the fingers and its sphygmogram give a steep ascent and pointed summit; yet it quickly vanishes to the fingers, and, as shown in the trace, it lacks duration, as compared with the duration of the systole which has sent it forth. If the heart passes into irregular action there will be systoles without arterial pulsation, and pulsations presenting a great variety in form and in time. The many and grave ulterior pathological sequences will not be followed, as the immediate mechanism is all that concerns the present study.

If the above sketch of the mechanism of mitral insufficiency be truly drawn, the conclusion inevitably follows that abnormal delay of the pulse is an essential phenomenon of the lesion.

Experiments on the Schema in Illustration and Evidence.

Passing next to the results of experiment, there is first offered those obtained with the schema. It is well known that the mechanical action of the circulation can be well imitated by means of a good working schema. With such a device connected with the transmission tubes of an apparatus for simultaneous inscriptions, traces were obtained at the same time from the interior of the ventricle and interior of the aorta. The intra-ventricular pressure in diastole was maintained at four inches, and the intra-aortic at twenty inches (water manometer). The ventricle was compressed and decompressed in a uniform manner for all the experiments. The conditions were varied only with respect to the mitral valve. The instant of ventricular impulsion and that of the aortic wave were signalled by the ascents of the traces, and the time difference between these ascents was easily computed by the accompanying chronogram. Experiments were first made with the mitral valve intact, representing normal conditions. No. 1, Fig. 101, is a fair example of the results. Delay of the aortic wave, $\frac{1}{16}$ second.

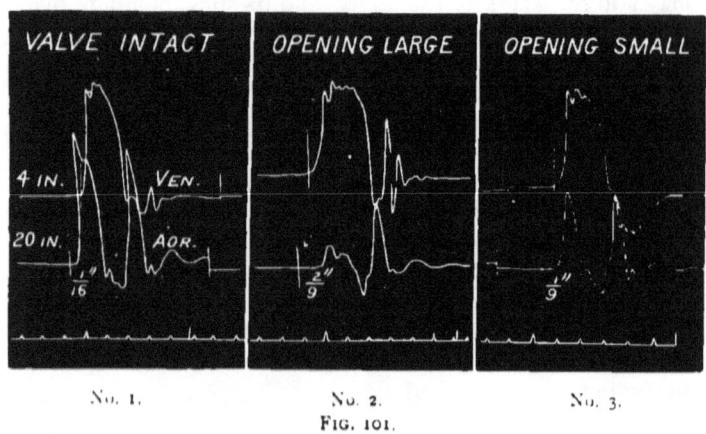

No. 1.　　　　No. 2.　　　　No. 3.
Fig. 101.

Another series of experiments was made with a large mitral opening, of which No. 2 shows an average result. Aortic delay, $\frac{2}{9}$ second—more than three times as long as the result with intact valve.

A third series of experiments was made with a comparatively small mitral opening, with result shown in No. 3.

Aortic delay, $\frac{1}{9}$ second, which is seen to be intermediate between the other two.

These results well illustrate the retarding effect of mitral insufficiency on the time of the pulse-wave, and the additional significant fact that the retardation is in proportion to the size of the mitral opening. Indeed, in view of the careful manner in which the apparatus was prepared and the experiments conducted, together with the analogy in all essential points for our purpose between the schema and heart, these results may be admitted, also, as veritable demonstrations of the phenomenon and phases under consideration.

Experiments on Man.

The subject remains to be further elucidated by the results of experimental observations on man.

In order to appreciate chronometric discrepancies between the heart and pulse, it is essential that the normal cardio-arterial time-relations should be well established. This has been done. The interval between the beginning of ventricular systole and the carotid pulse has been found, after a great many observations and measurements, to average about $\frac{1}{13}$ second with pulse rate at 72. The normal range of variation with this pulse rate may be placed between $\frac{1}{11}$ and $\frac{1}{13}$ second. It has been determined that the interval bears a relation to the pulse rate, so that a frequency of 120 would give an interval perhaps not more than half as long as a frequency of 60 per minute. Also, investigations have shown that the causes of the variations with the pulse rate are changes in the mode of ventricular systole and variations of the arterial blood pressure relative to the ventricular. In frequent pulsations, the systoles being quicker, and the arterial pressure relatively lower, the cardio-carotid interval would be correspondingly shortened. The rule is this: the cardio-carotid interval for a pulse frequency between 60 and 120 varies inversely with the pulse rate; so in estimating what would be a normal interval in a given case, the rate of the pulse must always be taken into account.

The interval between the heart and more distant arterial points, as the radial, femoral, or posterior tibial is complicated by the element of transmission time of the pulse-wave, which is also subject to variations; but the transmission interval between the aortic orifice and carotid point is so short, compared with the whole cardio-carotid interval, as measured, that its variations are not appreciably apparent. Hence the importance of choosing the carotid or subclavian for observing the cardio-arterial time interval.

The cardio-carotid interval, then, is practically the presphygmic, or time elapsing between the beginning of ventricular contraction and opening of the aortic valves, and variations of delay of the carotid pulse are virtually variations in the length of the presphygmic interval.

With these data concerning the physiological duration and variations of the cardio-carotid interval we are prepared to signal and appreciate the changes effected by conditions of disease.

Our immediate study concerns delay of the pulse in mitral insufficiency. The only clinical data we possess relating to this point have been furnished by the author's researches.

The first observation was made in July, 1877, on a boy, aged nine years, who suffered from mitral insufficiency, and subsequently dying, a post-mortem revealed in his case a pure mitral regurgitant lesion without any other valvular change. His pulse at the time of observation was 100 per minute. Simultaneous traces of his heart and carotid gave an interval of $\frac{2}{17}$ second, when about $\frac{1}{17}$ second would have been his normal interval with his pulse rate. Also simultaneous traces of his heart and radial gave an interval of a little less than $\frac{1}{5}$ second; and the same of the heart and posterior tibial a little more than $\frac{1}{4}$ second; both the latter also abnormally long. These results are shown in Fig. 102.

Soon after this case another was traced; a young man, aged twenty years, who presented all the diagnostic signs of mitral insufficiency. Simultaneous traces were had of his

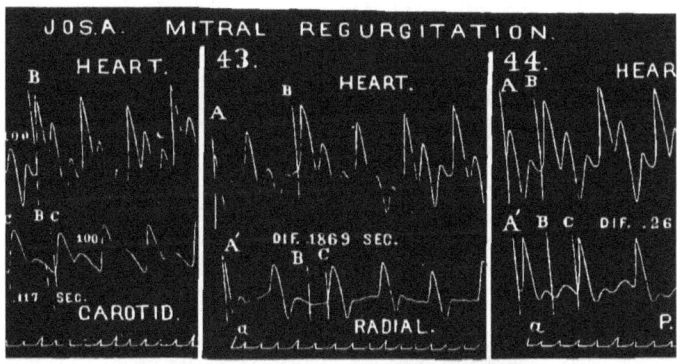

FIG. 102.

heart and radial, which gave an interval of $\frac{2}{5}$ second, with pulse rate at 80. If we deduct from this interval $\frac{1}{15}$ second, as an average carotid-radial transmission time, we get a value between $\frac{1}{5}$ and $\frac{1}{4}$ second as the approximate cardio-carotid interval in this case. The normal interval with pulse at 80 would be near $\frac{1}{18}$ second. Traces shown in Fig. 103.

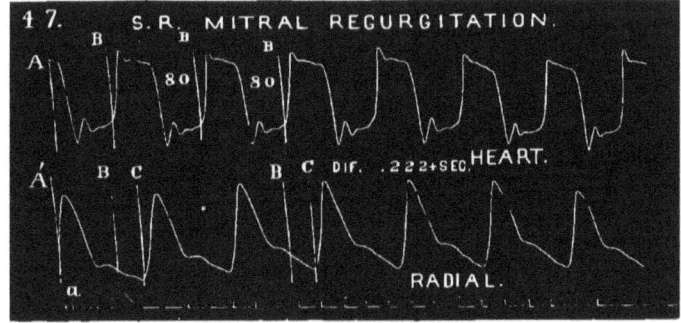

FIG. 103.

Since these two cases, observations have been continued with results which only multiply the demonstrations of the fact that the pulse is abnormally delayed in mitral regurgitation. But as demonstrations are more conclusive than mere statements of results, the simultaneous cardio-arterial traces (with accompanying chronogram in fifths of seconds), from a number of patients presenting diagnostic signs of mitral regurgitant lesion are here produced.

The cardio-arterial intervals in all these cases of mitral insufficiency are abnormally great; greater than those obtained from persons in health, or from those affected with any process of disease which does not involve pulse-retarding lesion of the heart or aorta. In the absence of such lesion, no series of cases with corresponding pulse frequencies can be produced which will show such an amount of pulse retardation.

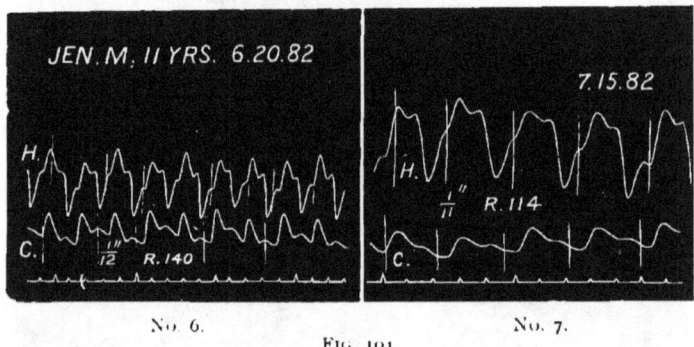

No. 6. No. 7.
FIG. 104.

Fig. 104, Nos. 6 and 7, are from Jennie M., aged eleven years. She had suffered with chorea and recurring rheumatism, and her heart had become seriously affected. It was enlarged, and emitted a loud murmur, which was distinguished as mitral regurgitant. No. 6 was taken during an acute attack, in the presence of fever, pulse rate 140. No. 7 was taken after subsidence of acute symptoms, pulse rate 114.

Some months subsequently the girl experienced a new attack, with grave cardiac implication, under which she died.

The post-mortem revealed pericarditis old and new, with adhesions and liquid effusion; general enlargment of the heart, especially hypertrophy of the right side; notable insufficiency of the mitral valve, the result of inflammatory thickening and contraction; the other valves normal.

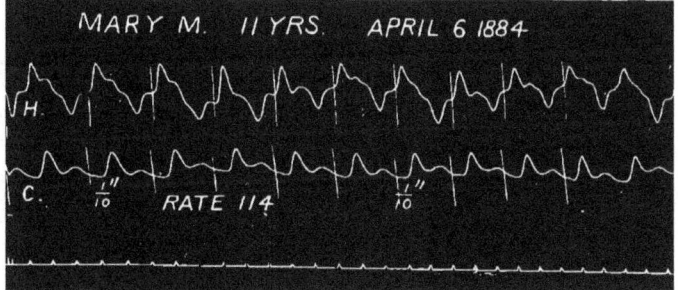

Fig. 105.

Fig. 105 is from Mary M., next younger sister of Jennie, also aged eleven years at the time of observation. She became affected with chorea and subacute rheumatic symptoms, and when first seen her heart was notably enlarged, and there was present a distinct apex systolic murmur. She had slight fever, pulse rate 114. Carotid delay shows $\frac{1}{10}$ second against $\frac{1}{19}$, which would be about her normal interval.

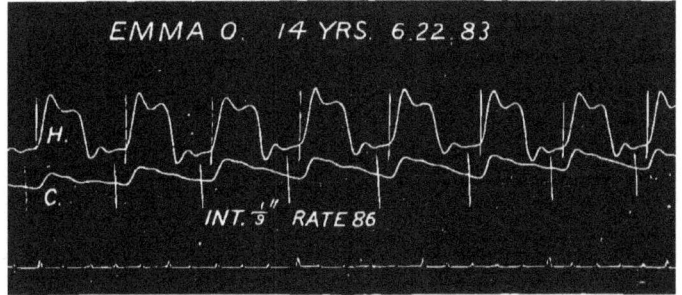

Fig. 106.

Fig. 106 is from Emma O., who presented an apex systolic murmur, which supervened during an attack of scarlet fever, and persisted after recovery from that disease. At the time of observation, June 22, 1883, the murmur and pulse delay were the only indications of cardiac lesion. She remains well to the present time.

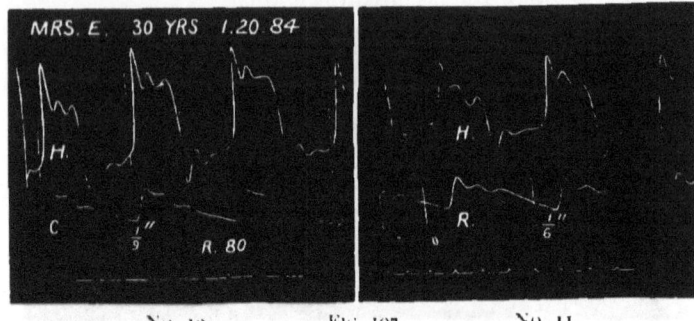

No. 10.   Fig. 107.   No. 11.

Fig. 107, Nos. 10 and 11, are from Mrs. E., aged thirty years. She was taken with præcordial soreness and a dry, hacking cough, and, when seen a few days afterwards, betrayed a marked apex systolic murmur. Tracings show the cardio-carotid interval to be $\frac{1}{9}$ second with pulse rate at 80, and the cardio-radial interval to be $\frac{1}{6}$ second.

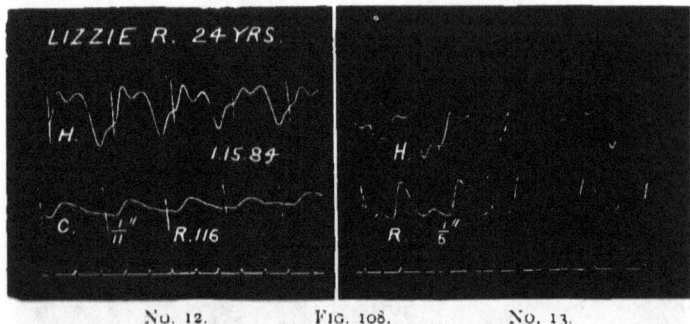

No. 12.   Fig. 108.   No. 13.

Fig. 108, Nos. 12 and 13, are from Lizzie R., aged twenty-four years, hospital case, presenting apex systolic murmur. Delay as shown.

Fig. 109, No. 14, is from Geo. B., aged twenty-seven years, presenting apex systolic murmur. Delay as shown.

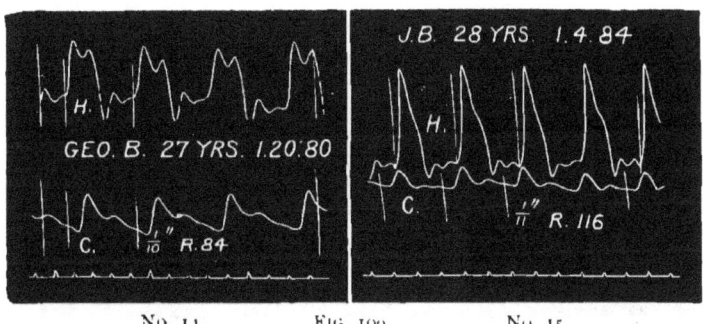

No. 14.   Fig. 109.   No. 15.

Fig. 109, No. 15, is from J. B., aged twenty-eight years, hospital case, with diagnosis of mitral regurgitation and cardiac hypertrophy. At the times the traces were taken he was under a febrile paroxysm. Indications and delay as shown.

Fig. 110, No. 16, is from J. T., aged fifty years, a hospital patient, with cardiac enlargement and apex systolic murmur. Pulse delay and pulse rate as shown.

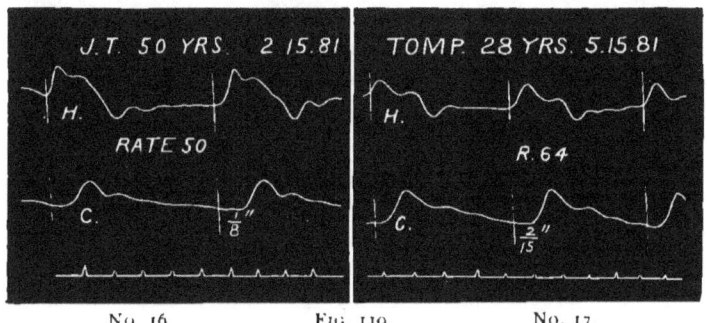

No. 16.   Fig. 110.   No. 17.

No. 17 is from Sam T., aged twenty-eight years, who presented the signs of mitral insufficiency.

Fig. 111, No. 18, is from J. C., a hospital case, with well-marked signs and symptoms of cardiac hypertrophy and mitral regurgitation.

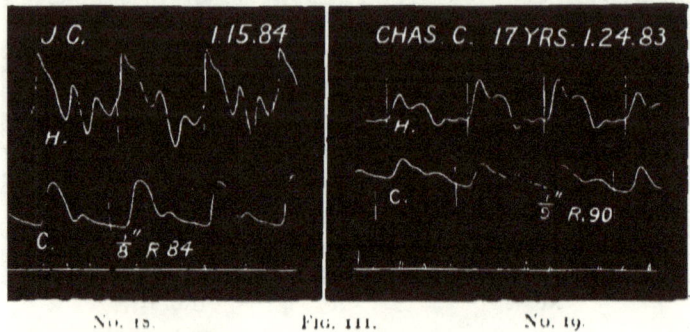

Fig. 111.

No. 19 is from Chas. C., aged eighteen years, with indications of mitral insufficiency.

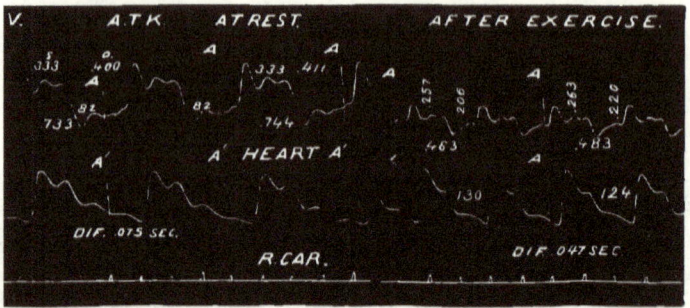

Fig. 112.

Fig. 112 is from a man in health, aged fifty-four years. The first traces were taken when the circulation was quiet, pulse 82; the second traces when the circulation was accelerated as consequence of exercise, pulse 130. In this figure the pulse delay represents the normal amount. It measures, in fractions, a little less than $\frac{1}{13}$ second, with pulse at 82, and a little less than $\frac{1}{21}$ second, with pulse at 130.

Having thus shown that the mechanism of mitral insufficiency includes increased delay of the pulse as one of its essential phenomena, and demonstrated the constant relationship between the lesion and phenomenon, both on the schema and man, the proposition is submitted as proven, and the fact as true beyond a doubt.

Diagnostic Value.

In undue retardation of the arterial pulse on the systole of the ventricle we have a special and highly diagnostic sign in mitral insufficiency. In this lesion, when pure and effective, the sign is always present, and fails to manifest only when the lesion is so slight as to cause no functional disturbance, or when complicated with large insufficiency of the aortic valves. In the condition of harmless regurgitation through the mitral valve the retardation of the pulse remains normal, because the conditions are virtually normal; in the complication stated the retardation is abnormally diminished, for the reason that the aorta, ventricle, and auricle are as one cavity, with blood pressure in equilibrium at the end of ventricular diastole, and the beginning of systole starts a current forwards into the artery as well as backwards into the auricle. Thus in mitral, complicated with free aortic, insufficiency, instead of abnormal delay, there is abnormal precipitation of the pulse. These are the only exceptions to the rule, that the sign is present when the lesion is present.

What other conditions besides the prime one considered are capable of determining undue retardation of the pulse? Slowness of ventricular contraction, and relatively high arterial pressure are known causes of prolongation of the presphygmic interval, but these causes are considered to produce only the normal extreme of prolongation.

1. There are good reasons to believe that mitral stenosis independent of insufficiency is capable of causing a delay which may exceed the normal limits. The mechanism, and

experiments on the schema favor this view; besides, not long since, a case was traced presenting the presystolic murmur and other clinical features of pure mitral stenosis and the cardio-carotid interval found unduly long. If this fact should become established, it will be found that the delay, unless in extreme narrowing, will not be as great as in mitral regurgitation, and that the sign will not manifest at all in moderate or even considerable contraction, for it would require a very small opening to prevent the ventricle from duly filling during diastole, which would be requisite to insure the production of the sign. Even admitting the fact of the relationship, the sign can have nothing like the diagnostic significance in mitral constriction that it has in mitral regurgitation.

2. Rarely, increased delay may be produced by a "locking" of pathologically changed aortic valves. This condition is uncommon, but may obtain when the valves are stiff and thickened, and, notably, when loaded with calcareous masses. They close with a species of locking and are forced open only after a prolonged effort of the ventricle.

3. It has been well established that an intervening thin-walled distensible aneurism of the ascending aorta may cause increased delay of the distal pulse, but that if the aneurism have resisting walls, even though it be large, there is no increased delay. Evidently, if the pulse could be traced on the proximal side of the pouch, which is impracticable in man, the delay would not obtain.

The conditions have now all been named in which there is a possibility of undue delay of the pulse being produced by other causes than mitral insufficiency. With the latter all are serious organic lesions of which abnormal delay of the pulse may be a common product.

The sign then, although always present in pure, effectual mitral insufficiency, is not of itself pathognomonic of this lesion, since it may be present, also, in other organic conditions. Here the stethoscope comes beautifully to the aid of the graphic instrument, and in the union of the apex systolic murmur of the one and the pulse-delay of the other, our diagnosis is perfected.

The graphic sign compared with apex systolic murmur.— Systolic murmur located at the left apex is the physical sign above all upon which turns the diagnosis of mitral insufficiency. Yet all know there may be free regurgitation without murmur, and insignificant harmless regurgitation with loud murmur, and that no direct relation subsists between the amount of regurgitation and the intensity of this auscultatory sign. Also is it well-known that apex systolic murmur may be distinctly present and the mitral valve be functionally perfect. These points do not compare altogether favorably with the point we have stated as true concerning the graphic signs. The comparison is well shown by apposition.

Abnormal pulse delay.	*Apex systolic murmur.*
Always present in pure, harmful mitral regurgitation.	Sometimes absent in very free regurgitation.
Always absent in harmless regurgitation.	Usually present and often very loud in harmless regurgitation.
The amount of delay directly proportional to the amount of regurgitation.	No certainty that the intensity of the murmur is directly proportional to the amount of the regurgitation, but often it is inversely so.
Present in other conditions besides mitral regurgitation, but these are organic and serious.	Present in other conditions besides mitral regurgitation, but these, if organic, are harmless.

Although the graphic sign has greatly the advantage in the comparison, its value and utility are most manifest when it is obtained and considered in connection with the auscultatory. Alone it is not competent to declare the presence of the lesion, but in association with systolic murmur at the left apex it speaks with positive import. Neither sign singly is conclusive of the lesion, but the two combined could be produced by no other. One locates the lesion, the other determines the extent of its damage.

An important point is that the diagnosis as made by the aid of the graphic sign is certain and immediate at any stage of the case, while the diagnosis as made by all other indications, exclusive of this, is not certain until confirmed by the development of pathological sequences. When, as often happens, other signs raise the question of the existence or not of harmful regurgitation, the graphic comes in with its prompt and positive decision.

Résumé of Conclusions.

1. Abnormal retardation of the arterial pulse, notably the carotid, on the systole of the ventricle is a real phenomenon of mitral insufficiency.
2. It is present in all cases of pure, harmful mitral insufficiency, and is absent only in insignificant, harmless regurgitation, or in regurgitation complicated with aortic insufficiency.
3. It measures by the amount of retardation the amount of the regurgitation.
4. It may be present in two other conditions, probably a third, all organic; hence this sign, notwithstanding its positive value, is not of itself pathognomonic.
5. In conjunction with apex systolic murmur its presence is conclusive of mitral insufficiency.
6. Compared with apex systolic murmur, this sign is more positive and appreciative, and distinguishes, which the latter does not, between harmful regurgitation and harmless conditions.
7. By the aid of this sign a positive diagnosis may be made at once in any stage of the case, and without waiting for the development of sequences and symptoms.

Enormous Retardation of the Pulse from Mitral Insufficiency, Aortic Aneurism, and Heavy Aortic Valves.

Exaggerated delay of the pulse has already been demonstrated respectively in mitral insufficiency, aortic obstruction in the form of heavy aortic valves, and aortic aneurism.

The case now to be presented gave an enormous delay of the pulse, and the post-mortem revealed a combination of the three pathological conditions just named.

The case was in the Cincinnati Hospital, and obligations are due to the physicians in charge for the opportunity to observe it.

Chris. H——, aged thirty-one years, presented a systolic murmur, distinct over the front of the chest and heard in the back. The murmur was accentuated both at the apex and second left interspace. Second sound clear and loud. Graphic records were taken September 6, 1879. Simultaneous tracings, with accompanying chronogram, were obtained of the heart and left subclavian, of the heart and right carotid, of the heart and right radial, and of the heart and left radial. These are all shown in Figs. 113 and 114, except the heart-carotid tracings.

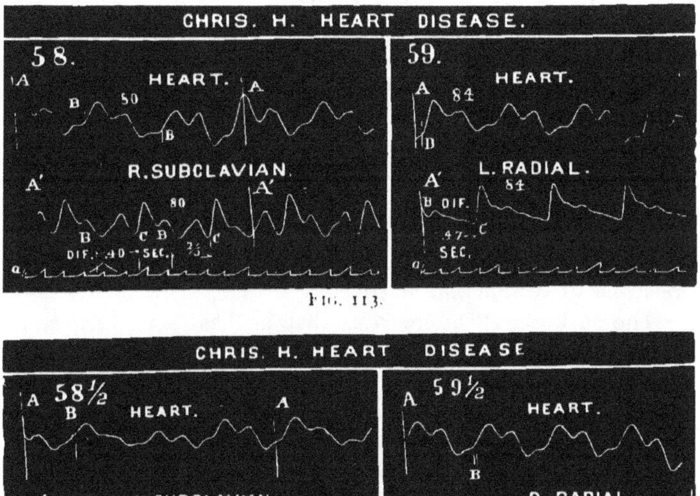

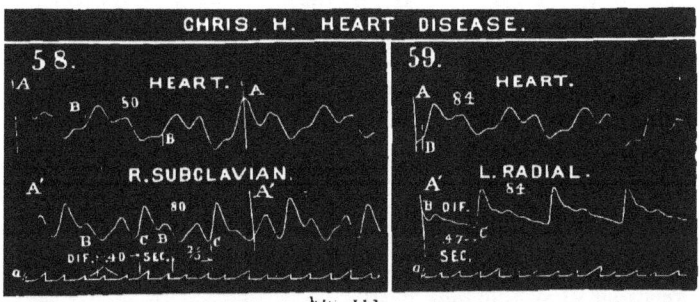

Fig. 113.

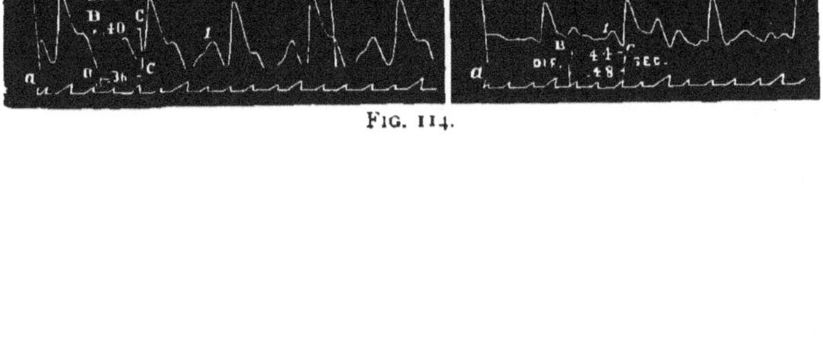

Fig. 114.

Patient died November 16, 1879, and here are notes of the post-mortem, taken from the hospital case-book: Heart weighed fifteen ounces, and was fatty. Just above the aortic valves were two aneurismal pouches, one projecting forward and to the right, the other forward and to the left. The first of these, looking like dilatation of the *sinus* of *Valsalva*, contained no clots; the other was filled with laminated fibrine. The latter was the size of a small apple, and presented a distinct ring about half the diameter of the pouch. It pushed its way in various directions, and pressed upon the pulmonary artery in such manner as to occlude this vessel entirely. Both dilatations had very thin walls. The aortic valves were much thickened, but were competent by the hydrostatic test. The mitral valve was filled with nodular vegetations, and was incompetent. Right auricle distended and filled with soft clot and fluid. Tricuspid and pulmonary valves presented no disease. Pericardium thickened, and contained twenty-one ounces of fluid. Aorta atheromatous.

Recurring to the tracings, the cardiograms fairly show the phases of the cardiac revolutions, and the beginning of systole cannot be mistaken except in a very insignificant sense. The line B in the figures must indicate very closely this beginning. As to the end of systole, it is located near midway of the main descending line. The first pulsation after B No. $58\frac{1}{2}$, Fig. 114, probably best shows the time-relations of systole and diastole.

The author well appreciates that he has located the end of systole at the highest point of the main descent, and still believes that this localization will hold in most cases of perfect traces of normal hearts; but in such traces as we are able to get of pathological hearts, often the line begins to fall, in consequence of depletion of the ventricle, before the end of systole is reached. The change then transpires in the line of fall, and usually impresses there no definite mark.

According, then, to the interpretation given the cardiograms in this case, systole, as compared with diastole, is excessively long.

Respecting the arterial traces, the form of both subclavians is *unique*. (The carotid shows essentially the same form.) The peculiarity is carried to some extent into the right radial, but is lost in the left. This singularity of form will not fail to arrest the reader's attention, and its explanation will be sought farther on. A point worthy of notice is that a just interpretation of these traces will show that the cardiac systolic portion of the pulse is disproportionately short.

But the great lesson of the case concerns the extraordinary delay of the pulse on the systole of the ventricle. Very careful measurements give, as shown (Nos. 58 and 59), $\frac{40}{100}$ second as the delay of the right subclavian, and $\frac{47}{100}$ second as the delay of the left radial. The enormity of this delay is apparent when it is remembered that the normal delay of the subclavian, with pulse at 80, would not exceed $\frac{1}{13}$ second. It is believed that the retardation of the pulse in this case is the greatest ever observed—certainly the greatest that has been recorded; and if the post-mortem had not furnished the remarkable combination of lesions to explain the delay, it would have been difficult to disarm the criticism that the phenomenon was fallacious and existed only in an error of observation. The three great causes of pulse-retardation coexisted in the case, and the graphic method was competent to demonstrate the amount of retardation so produced.

The physical signs pointed to a cause of systolic murmur located at the pulmonary artery, while the second centre of accentuation of the systolic murmur at the apex indicated, so far, mitral insufficiency; but it was easy to conceive that this murmur might be propagated from the base. As affection of the right side could only produce negative graphic results as regards the pulsations of the heart and arteries, the very positive graphic showings obtained demonstrated the existence of grave disease located at the left side. What manner of disease affecting the left side of the heart could produce the graphic phenomena presented? As to the strange pulse-form, this was an enigma. It was passed without explanation. As to the pulse-delay, the organic conditions producing this phenomenon having been previously ascertained, it was plain that the delay in question must have arisen from one or more of these causes. The limits of pulse-retardation under its causes in single or combined action not having been determined, it was impossible to know with certainly whether such an amount of delay would require for its production all the pulse-retarding lesions; or whether two, or even one, might be sufficient. The author's observations had shown that the delay is great in mitral insufficiency, enormous in conjoined mitral insufficiency and heavy aortic valves, and distinctly notable in aortic aneurism. From the graphic data, in connection with the physical signs, the diagnosis was reached that the patient was suffering certainly from mitral insufficiency, and probably from added aortic obstruction in the form of heavy valves; but only vaguely did the idea of associated aortic aneurism occur. Illustrated now by this case, such extreme delay of the pulse would be accepted as diagnostic of the three important pathological conditions under consideration.

The existence of aortic aneurism in association with the other lesions afforded an explanation of the singular pulse-form. The ventricles in their systole compressed the overlying aneurisms, causing the latter to discharge a portion of their blood into the aorta proper, which initiated a considerable wave. This wave, marked 1, was transmitted so much in advance of the true discharge-wave from the heart, because escape of blood from the ventricle was delayed by the existing cardiac lesions. The deviation in the upward slope of this wave was probably made by the auricular systole, possibly by default of synchronism between the two ventricles. The ventricle in the latter part of its systole threw quickly a comparatively small amount of blood into the aorta, which accounts for the blending of the second with the early appearing aortic wave.

This description of pulse, it would seem, could only be reproduced by a similar conjunction of conditions, and its demonstration would constitute an important diagnostic sign of these combined lesions.

Thus, step by step, as correct graphic phenomena are connected with well-ascertained pathological conditions, will the science of clinical sphygmography be built up.

CHAPTER IX.

CARDIO-SPHYGMOGRAPHIC HISTORY OF AORTIC OBSTRUC-
TIVE LESIONS—THE SPHYGMOGRAPHIC INDICATIONS
IN ANEURISM.

MRS. T——, aged sixty years, suffered for several years from organic heart disease. She was thin and incapable of active exercise, but by going carefully, usually attended to her ordinary household duties. Well-marked signs of hypertrophy of left ventricle. Action of heart strong, with jogging impulse and regular rhythm. Systolic murmur, greatly emphasized at second right interspace, where also slight undulation and thrill are perceived. The murmur propagated upward along course of aorta, faint in other directions; at apex audible, but faint; audible in back. First sound well heard at apex, but supplanted at base by murmur. Second sound distinct. No diastolic murmur. Radial pulse to fingers firm, slow, and prolonged. The clinical history admits of only one diagnosis: aortic obstruction without aortic reflux, and coexisting hypertrophy of the left ventricle. The tracings shown were obtained at one sitting, July 16, 1878. The case, though evidently not one of extreme obstruction, from its well-marked features is well suited for sphygmographic study.

The cardiograms present an individuality of form which at once arrests attention. The systole is sustained as if the ventricle were laboring against an obstacle, and is slowly changed into diastole as if its work were not completely finished. The most striking feature is the great prominence of the auricular wave. One interpretation only can be given this phenomenon, viz., that it denotes hypertrophy of the left auricle. The tracings alone reveal this important feature of the case.

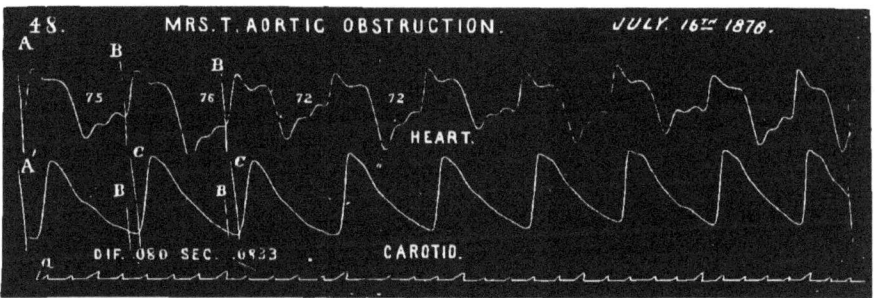

Fig. 115.

The tracings of the carotid pulse, Fig. 115, and of the radial, Fig. 116, are characterized by sloping ascents, rounded tops, and imperfect delineations of the secondary waves. They show also high amplitude—notably of the carotid

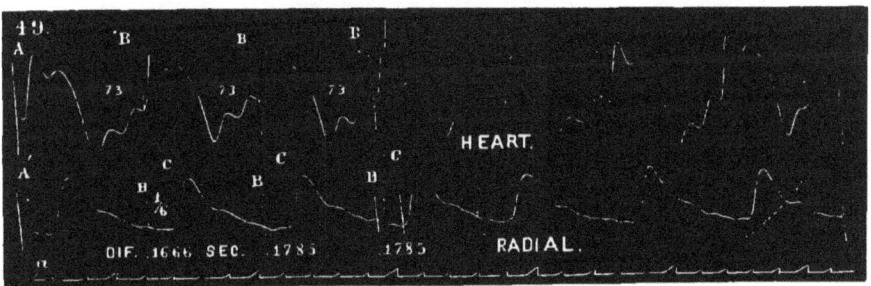

Fig. 116.

pulse—and high arterial tension. So while the first features indicate impediment to the exit of blood from the ventricle, the latter demonstrate that the arteries notwithstanding are well charged at each systole.

The time difference, as shown, between the heart and beginning of the carotid pulsation is .0833 = one twelfth of a second, and that between the heart and beginning of the radial is .1666 = one sixth of a second; which are correctly normal. But it will be observed that the interval between the heart and summit of the pulse is much greater than normal.

The type of pulse-trace instanced is constant in material aortic constriction, and the degree of deformation marks the amount of impediment to the passage of the blood. The type, however, may be simulated in the pulse below an aneurism or other source of arterial obstruction located beyond the aortic root.

The sphygmographic indications of ordinary aortic obstruction may be formulated thus: Heart's pulsation with sustained systole; arterial pulsation with sloping ascent and rounded or flattened top; interval between beginning of cardiac systole and beginning of arterial pulse normal.

A remarkable showing in this case remains to be considered. In Fig. 117, while the cardiogram conforms to the

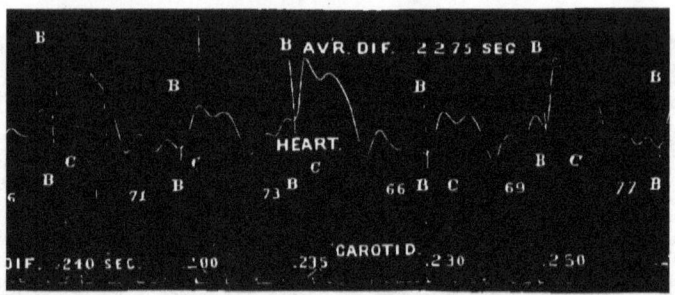

FIG. 117.

others, except in the greater modification induced by the respiration, the carotid trace presents a wholly different type from that of Fig. 115; and especially will it be observed that the carotid follows the cardiac pulsation at the very long average interval of .2275 second, which is about three times the time difference of the other record. This is the only registry of the kind obtained among several observations, the others agreeing closely with Fig. 115. There was no fallacy. What is the explanation of this extraordinary change and contrast? It has been demonstrated that mitral regurgitation causes abnormal delay of the pulse. It is not difficult to conceive that under favoring conditions an *intermittent* reflux through the mitral valve may take place; and that in this instance, under the action of the hypertrophied ventricle and the mean-time sufficiency of the mitral valve, the ventriculo-arterial blood pressure is at intervals increased to the point of preventing closure of the mitral segments, which then permit of reflow until the blood pressure in front is reduced to the status at which the valve again becomes competent. Under the conditions, the behavior of the mitral valve would be analogous to the safety-valve-action claimed for the tricuspid.

The alternative explanation applies a theory before advanced. It is that the changed aortic valves, although in this case usually yielding with sufficient promptness, at times become *fixed*, so to speak, and require additional time and force to open them. However, it is far easier to conceive of an intermittent patency of the mitral valve than of an intermittent fixing of the aortic valves. The mitral regurgitant theory also receives support from the existing auricular hypertrophy; for although this condition arises as a more remote effect of aortic obstruction, it is the natural and direct result of mitral reflex. Again, the peculiar and striking change of form of the carotid pulse can be accounted for better on the supposition of mitral insufficiency than on that of aortic fixing. On the other hand, in favor of the alternative theory, is the positive evidence, aside from the sphygmographic, of the existence of aortic lesion, and the negative evidence, save the sphygmographic, of the supervention of mitral reflux. It may be, however, that an apex systolic murmur, had it been listened for, would have been heard while the peculiar record was being made. On the whole, it is best to accept the theory of intermittent mitral regurgitation in explanation of the recorded change in form and time of the carotid pulse.

The following are the notes of the examination, January 20, 1881: Action of heart vigorous, but less jogging than formerly; loud rasping basic murmur, systolic in time, with marked accentuation at second right interspace, where are

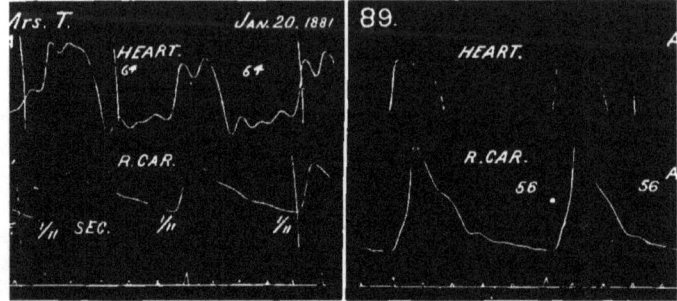

Fig. 118.

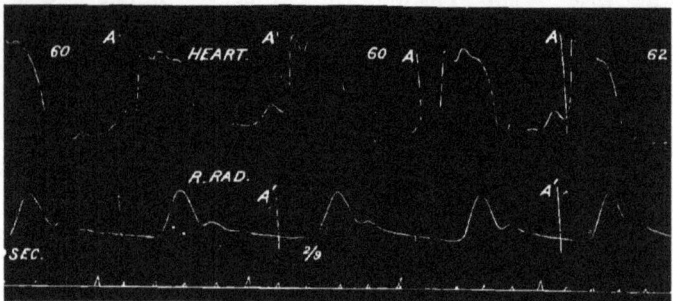

Fig. 119.

perceived also a feeble impulse and thrill. Second sound of the heart at aortic site covered by a feeble, short murmur; at pulmonic site, normal. Heart changes its rhythm from forty to seventy per minute; at times its action is intermittent, impulse strongest in fifth interspace outside the nipple line. Pulse prolonged; varies markedly at intervals in amplitude, force, and rhythm. Graphic observations were taken simultaneously, in repetition, of the heart and carotid, and of the heart and radial arteries, of which results are shown in Fig. 118, Nos. 88 and 89, and Fig. 119.

The reader is asked to carefully compare the above traces with the former ones.

The autopsy was conducted by Dr. A. B. Isham, assisted by Dr. Lowry and the author. These are the notes: General anasarca and ascites, pericardium filled with serum, heart much enlarged, all its cavities and great veins distended with blood; left ventricle greatly hypertrophied, walls an inch thick; left auricle not enlarged; right auricle dilated and hypertrophied; right ventricle dilated and hypertrophied; aorta and pulmonary artery both dilated; the mitral, tricuspid, and pulmonary valves normal, except expanded proportionate to the cardiac hypertrophy; aorta in some degree atheromatous, yet presenting a smooth structure above the valves; aortic valves the seat of extraordinary changes. Of normal extent, they carried on their upper surfaces enormous masses of rough calcareous matter. Two were united by their borders for nearly half their extent, leaving a triangular space, and forming a rigid septum, which contracted the arterial opening to about one third the normal size. The third valve, though heavily loaded, was pliable at its attached border, and performed the office of opening and closing this contracted orifice. The calcareous mass on this valve measured three eighths of an inch through its thickest part; the sinus behind was dilated so that it had no room to play. Of the conjoined valves, the calcareous mass over one filled up the space between it and the aorta, and measured five eighths of an inch through its middle; the mass over the other was disposed in chains extending between it and the aorta. The under surfaces of the valves were smooth, except opposite the attached borders, where calcareous tubercles presented.

This valvular disposition included a most remarkable feature. The third, or functioning valve, would move on its attachment, rising and falling *en masse*. Ordinarily, when down, it overlapped by a little the rigid border of the opening; but when pressed from above, it would sink below the rim, and remain there as if locked, until pressed from below, when it would rise with a spring.

In the light of the sequel and post-mortem developments, the case can now be reviewed with peculiar interest and profit. The second graphic examination gave results in consonance with those of the first, although separated by an interval of two and a half years; so it was concluded that the diagnosis first arrived at was still sustained. The only real difference in the physical signs was the development of diastolic murmur over the aortic site, but this did not indicate the existence of aortic insufficiency to an extent worthy the name. The tracings showed the beginning of the arterial pulse in normal or delayed time, whereas had the valves been patulous the pulse would have begun distinctly earlier than normal. The autopsy proved this item of the diagnosis correct. Undoubtedly the murmur was due to the friction of the refluent blood against the rough deposits during the time the clumsy valve was falling into position. The form of the traces plainly indicated the presence of aortic stenosis, but not in an extreme degree, for the pulse was shown to be at times ample and tense. And this was the exact state of the orifice as proved *post-mortem*. The traces unmistakably indicated the presence of auricular hypertrophy, and this was confirmed *post-mortem*, but as respects the right auricle, instead of the left, as assumed. In this is the impressive lesson, that hypertrophy of the right auricle, as well as the left, may produce a high auricular wave. And in view of the coexisting right ventricular hypertrophy, it may be considered probable that the cardiac traces were produced entirely by the right heart. This view is supported by the strict conformity to the same type of the heart-traces, notwithstanding the marked changes in the pulse-traces. It may be remarked, however, that the ventricles always begin to contract synchronously, so the beginning of the ascent, in any event, marked the beginning of left ventricular systole.

The traces indicated the intermittent operation of some factor, which caused for the time a remarkable change in the time and form of the pulse. This change is strikingly shown in comparing Fig. 115 with Fig. 113, and has been fully described in the first account. The conditions of the aortic valves, found after death, afforded a full and complete explanation of this remarkable sphygmographic phenomenon; and it is interesting to note that the alternative explanations previously offered included the true one, and that the process there pictured was exactly that which in reality was taking place. The mechanism was this : Ordinarily the heavy valve which guarded the constricted orifice duly rose and fell, permitting the slowly rising pulse to begin within normal time after the beginning of ventricular systole. But at intervals, under vigorous strokes of the ventricle, and large volumes of blood sent forth into the aorta, in connection, perhaps, with impediment in the capillaries, the arterial blood pressure would become so enhanced that the valve in diastole would be forced below the rim and so become locked or fixed in position. The next systole would require unusual time to raise the intra-ventricular blood pressure to the point of dislodging the valve. When dislodged, however, the latter would rise with a spring, and would be carried wide open by the accumulated pressure behind, and the column of blood would enter with unusual quickness, to be cut short, however, by the termination of systole. Thus the pulse-trace would change in time from accustomed normal to great abnormal delay, and in form from accustomed slow ascent and rounded top to quick ascent and pointed top. After a few beats the usual *régime* would be restored, the change being brought about by the reduction of arterial pressure following the comparatively small volumes of blood sent into the artery at each systole.

In conclusion, the case well illustrates the rich and positive aids to cardiac diagnosis afforded by the graphic method. By this method alone could the amount of aortic stenosis have been so nearly approximated. By this method alone could the rare and peculiar form of aortic valvular obstruction have been in any sense made known. By this method alone could have been settled the question, raised by the presence of aortic diastolic murmur, as to competency or patency of the aortic valves. And by this method alone could have been determined the auricular hypertrophy, albeit it failed to distinguish the affected auricle. Besides the method joined the other signs in the determination of left ventricular hypertrophy, and gave a permanent record of the variations of the pulse in amplitude, celerity tension, and rhythm.

Sphygmographic Indications in Aneurism.

On the authority of a few distinguished names, supported, however, by a comparatively small number of experimental observations, it is currently taught that the pulsations below an aneurism afford two important indications of its presence : first, a characteristic form of pulse—a typical deformation ; second, an abnormal delay of the pulse. If these changes exist, the former can be demonstrated with the simple sphygmograph, and the latter with a very perfect differential sphygmograph; and if the question be thus simple and settled, as stated, we have indeed arrived, through the graphic method, at a wonderful and gratifying precision in the diagnosis of aneurism.

As regards deformation of the pulse below an aneurism, Marey* remarks: " When an aneurism exists on the tract of an artery, one observes below the tumor an important modification of the pulse; the quickness gives place to an extreme slowness, and often the touch is incapable of feeling this pulsation, because of the slowness with which the finger is raised," and represents the pulse below an aneurism by a tracing with slanting up-stroke, round top, and gradual descent without mark of secondary waves.

Dr. Mahomed† gives sphygmographic tracings of several cases of aneurism in which this imperfection is well shown.

* *La Méthode Graphique*, p. 582.
† *Medical Times and Gazette*, vol. ii., p. 141, 1873.

Of three cases of innominate aneurism verified *post-mortem*, one shows marked distortion of the right radial as compared with the left, while two present the right radial correct in form and uniform with the left. From his cases Dr. Mahomed infers that in pure innominate aneurism the right radial pulse presents the " aneurismal trace," while in aneurism of the innominate continuous with dilatation of the aorta, the right radial is little, if any, modified, and right and left correspond.

As regards delay of the pulse below an aneurism, François Franck* has contributed important observations. By means of an apparatus devised by himself he showed, in a case of innominate aneurism, that the right radial pulse was delayed on the heart $\frac{16}{100}$, and the left radial $\frac{11}{100}$ second; and in another case the right and left radial pulses were delayed on the pulsation of the tumor, respectively, $\frac{21}{100}$ and $\frac{14}{100}$ second. Thus François Franck esteems delay of the pulse below an aneurism as constant and of the highest diagnostic value.

CASE I. *Aneurism of the Innominate.*—For the opportunity of observing this case, indebtedness is acknowledged to Dr. Jos. Ransohoff, who also kindly furnished the notes of the *post-mortem*. Frank M——, aged forty-eight years, presents a prominent tumor at the site of the right sterno-clavicular articulation, which manifests pulsation, thrill, and a systolic murmur. Second sound of the heart clear and greatly intensified. Respiration much embarrassed by pressure of the tumor against the trachea. The graphic records were taken January 1, 1879. A few days subsequently the man died, and the autopsy revealed a large globular aneurism of the innominate artery, about three inches in diameter. The cavity contained fibrinous coagula, which encroached upon and obstructed the carotid orifice, but left the subclavian orifice free. The aneurism communicated with the aorta through an opening, only somewhat larger than the normal innominate lumen. The walls of the aneurism were

* *London Medical Record*, March 15, 1878; *Amer. Jour. Med. Sci.*, July, 1878, p. 258 ; *La Méthode Graphique*, p. 586.

thick and firm, and only showed some degree of thinning and yielding at the apex of the external tumor. The aorta was much dilated from its origin through the entire arch; aortic walls firm. The aortic valves were expanded and thickened, yet competent.

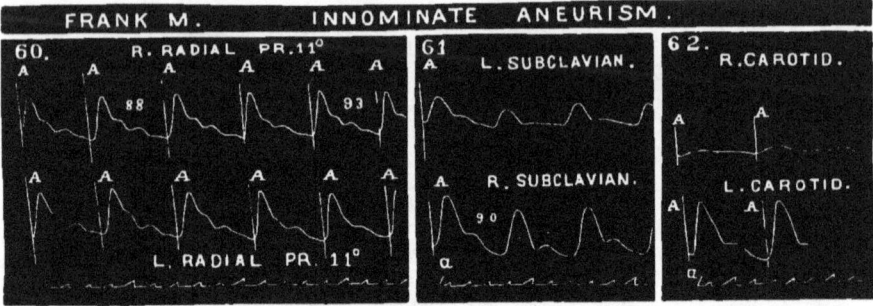

FIG. 120.

Of the tracings, No. 60, Fig. 120, shows the two radial pulses taken simultaneously. It will be observed that the pulsations are well formed—in fact, they show a rather low arterial pressure, but no departure from a normal type; that they are uniform with each other in amplitude, secondary waves, pressure degree, and all. Especially will it be noticed that the signal lines (purposely placed near the basal points) are precisely in the same relation to the basal points in both tracings, which demonstrates the exact synchronism of the right and left radial pulses. The same characters and conformity of the radial pulses are shown again in Nos. 64 and 65, Fig. 121. These representations also demonstrate that the time difference between the cardiac and radial pulsations is about one eighth of a second. (A seventh may be attained by extending the measurement beyond the lowest point of the radial traces, which is the more correct method for estimating the interval between the heart and arterial pulses.) Even a seventh of a second would not be greater than the average normal cardio-radial interval, with pulse at 90.

In No. 61, Fig. 120, owing to the more superficial position of the right subclavian, its pulse-trace is more ample than that of the left; the pulsations are shown to be synchronous.

In No. 62 the right carotid pulse compared with the left is greatly reduced and deformed, yet there is between the pulses no appreciable asynchronism.

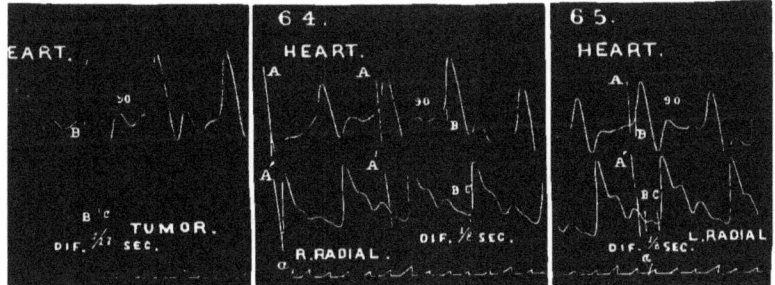

FIG.—121.

In No. 63 the *ligne d'ensemble* of the tumor-trace undulates in a marked manner with the respiration. In the ascending line, just after what at first view would be taken as the basal point, is an angle of deviation. Measurement from this latter and the basal point of the heart-trace gives $\frac{1}{17}$ of a second as the time difference between the pulsations of the heart and tumor—an interval entirely consistent; while measurement from the first point named and the cardiac basal point would make the heart-tumor interval extremely short—too short, in fact, to be admissible on any just theory as the true representation. Therefore, in this case, the second point is adopted as marking the beginning of the discharge-wave from the heart, which is indeed the true pulse-wave.

The sphygmographic records in the present case demonstrate:

1. That innominate aneurism of large size, communicating with a dilated aorta, may exist without the pulse beyond,

either in form or time, affording any evidence whatever of the presence of such aneurism; witness the correctness and perfect equality in form and time of the two radials.

2. That the effect of such aneurism on the *form* of the pulse beyond, may be positive and marked; witness the great inequality in form of the two carotids.

3. That the effect of such aneurism on the *time* of the pulse may be negative as to delay; witness the time between the heart and tumor, between the heart and radials, and the perfect equality in the time registry of the two radials, the two subclavians, and the two carotids.

In this case, also, the demonstration is complete, that the pulse-form was correct in the right radial, because of the free passage to the subclavian, and imperfect in the right carotid because of the obstructed passage to this artery. Hence the question arises whether the peculiar pulse-deformation, observed below some aneurisms, does not always depend upon arterial obstruction from filling of the lumen, or pressure of the tumor, or both. The question is one of importance.

The two following cases will aid our investigation. They were in the Cincinnati Hospital, and kindly placed at disposal for sphygmographic study. The notes were derived from the records and personal observation.

CASE II. *Aneurism of the Ascending Aorta.*—J. Bradford, aged 31 years; suffers from dyspnœa. Presents a prominent pulsating tumor a little to the left of the median line, and two and a half inches above the nipple. The tumor manifests thrill to the touch and murmur to the ear; the latter synchronous with the first sound of the heart, and heard over a considerable area. Tracings taken in August, 1877. Patient afterwards found his way into St. Mary's Hospital, where he died. The autopsy revealed a large aneurismal tumor, involving the ascending aorta from near the innominate downwards to near the origin, and extending from the upper border of the second to near the fourth rib; sternum eroded. On section, the walls for the most part were reinforced by thick layers of fibrine; although a number of accessory sacs presented with thin walls, one of which had ruptured into the right pleural cavity and proved the immediate cause of death. The left ventricle was somewhat enlarged and turned posteriorly.

Inscriptions of the heart's apex could not be obtained on account of its recedence from the chest-wall, therefore observations were limited to the tumor and arteries.

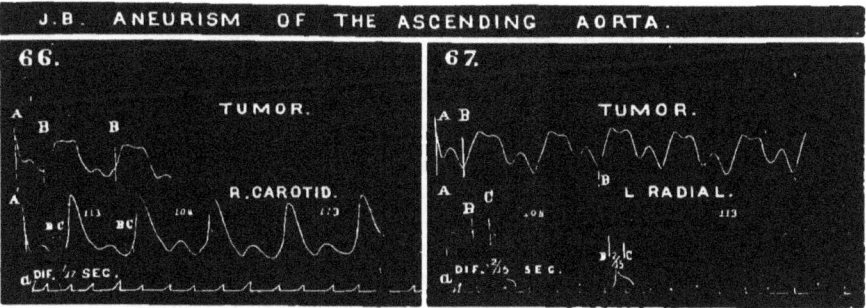

Fig. 122.

No. 66, Fig. 122, shows simultaneous tracings of the tumor and right carotid; No. 67 shows the same of the tumor and left radial. The pulse acceleration, (108–113,) the dicrotism, and low pressure will be noticed; also the characteristics of the tumor traces. The deviation in the lower part of the upstroke, sufficiently indicated in the tumor and carotid traces, but less distinct in the radial, probably marks in this case, as in the first, the beginning of the true pulse-wave, sent from the heart. The preceding lowest point, however, is more definite, and representing, as it unquestionably does, the same stage of the respective pulsations, is more certain and available for measurement. The interval between the pulsations in the tumor and carotid, measures, as shown, $\frac{1}{75}$ of a second; and that between the tumor and radial, $\frac{2}{75}$ of a second. If no cause of delay existed, the pulse-wave would pass over the distance represented by the tumor and carotid in $\frac{1}{35}$ of a second at the longest, and over the distance represented by the tumor and radial in $\frac{1}{15}$ of a second at the longest. The reason of the apparent greater delay of the carotid pulse compared with the radial is that the retardation takes place in the aneurism, and the pulse-wave reaching the free artery beyond, travels on with its accustomed velocity. In form, both the carotid and radial present the usual features of accelerated, moderately ample, and low-tension pulses.

CASE III. *Aneurism of the Ascending Aorta.*—Thos. Conley, aged thirty years, has suffered between two and three years from trouble within the chest. Front of chest, especially to left of sternum, bulging, and flat on percussion. Pulsation seen and feebly felt over an unusual area, more distinct than elsewhere in the fourth right interspace near the sternum. Absence of thrill, usually no murmur, although a systolic one has at times been detected. Patient very feeble; radial pulse small and frequent; suffers greatly from pain, dyspnœa, and cough. Died suddenly the last of February, 1879. The post-mortem disclosed an extensive aneurism of the ascending aorta, which had ruptured and discharged into the right pleural cavity. The aneurism commenced very near the aortic orifice, which latter was considerably dilated and involved the ascending portion of the vessel, which was expanded into a large flabby sac.

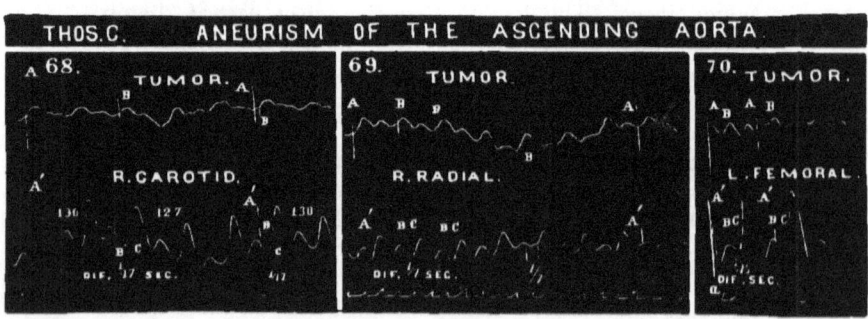

Observations were made January 31st. Fig. 123, Nos. 68, 69, and 70 are reproductions of the graphic records. The pulsations of the aneurism were taken at the fourth right interspace, but they were too feeble to produce good traces. However, the basal points are fairly discernible. The heart's pulsations could not be traced.

In the measurements, wherever there might be uncertainty as to the position of the basal points, those were selected which would afford the shortest interval; so if any error of representation exist, it is in making the time-differences shorter than the true ones. As shown, the tumor-carotid and tumor-radial intervals, respectively, are closely the same as in the preceding case. The tumor-femoral interval also shows prolongation, and that it is in proper relationship to the tumor-radial interval. The pulse frequency is 127-130, and the pulse-form reflects only a very common type.

Respecting the form of the pulse, these two latter cases support the first in opposing the idea that aneurisms, independent of arterial obstruction, impress upon the pulse beyond a peculiar deformation; the pulse-forms in these cases being only such as often present in other conditions. It seems reasonable, however, that there should be a description of aneurism with free arterial ingress and egress, which would cause the pulse-form of the affected distal artery to differ from that of the opposite unaffected artery, and yet not to the extent of a typical deformation.

In point of delay, these cases accord with François Franck's; and delay of the pulse, as an effect of certain aneurisms, may be accepted as a demonstrated fact. Can we differentiate the aneurisms which produce pulse-delay from those which do not? Case I., in which there was no delay, presented an aneurism of large size, free arterial passages, and *resisting* walls; while Cases II. and III., in which there was delay, presented aneurisms of large size, free arterial passages, and *yielding* walls. The conclusion would follow that the opposite conditions of the aneurismal walls produced the difference observed in the time of the pulses.

Observations are needed to determine whether a sacculated aneurism, communicating with the artery by a small opening, would cause delay of the distal pulse. It is conceivable that it would not. From what we have learned of the effect of large aneurisms on pulse-transmission, we are prepared to believe that small ones would cause no appreciable delay.

The author's observations indicate that the pulse-retardation in aneurism is much less than that in valvular disease of the heart, and greater than that of any general condition of the vessels and circulation.

Are there any other conditions than aneurism affecting an artery that may produce retardation of its pulse-wave? Two are suspected—namely, arterial obstruction, and the

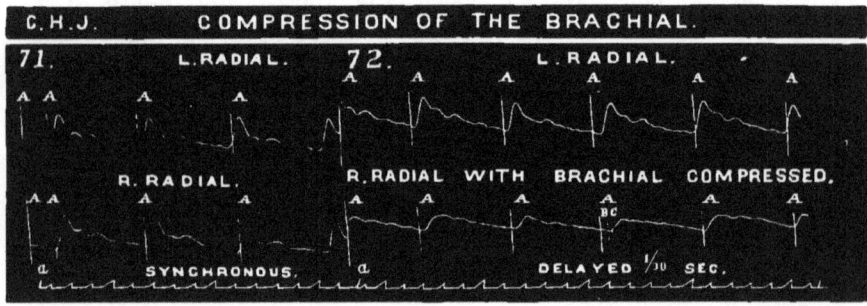

Fig. 124.

condition recognized as vaso-motor paralysis. These we will examine.

In Case I., Fig. 120, No. 62, the right carotid pulse, although greatly reduced and deformed by arterial obstruction, shows no appreciable delay. Yet it must be considered, that if delay arise from obstruction, it will be more evident at a distance from the site of obstruction; and that a point of observation so near as in this case could not be expected to show it, even if initiated. The effect of temporary compression of an artery on the velocity of its pulse-wave has been tested. Two healthy men were separately placed under experiment by first tracing their two radials simultaneously to prove them synchronous ; and after compressing one brachial so as to leave the pulse just perceptible, tracing the radials again, and noting the time-relationship of the two pulses. These observations many times repeated showed the pulse of the obstructed artery notably behind the other in point of time. Fig. 124, Nos. 71 and 72 give a fair representation of the results obtained.

The delay of the obstructed pulse compared with the other is about one thirtieth ($\frac{1}{30}$) of a second, which would about double the usual delay of the radial on the brachial pulse. This demonstration, however, is conclusive only as to the effect of temporary obstruction, and the query remains whether in permanent obstruction in which the artery would be contracted and adapted to the blood-stream, such retardation would occur. Data are wanting for the settlement of this point.

In one of François Franck's cases of innominate aneurism, the sphygmographic tracings show the right radial pulse fuller and better developed than the left, and the temperature was elevated 1° F. on the right side. These phenomena were attributed to vaso-motor paralysis, produced by pressure of the tumor on the first thoracic ganglion of the sympathetic. The influence of vaso-motor paralysis on pulse-wave velocity is an interesting question which presents for investigation.

The view has been advanced that the peripheral vascular dilatation and consequent easy flow of blood in the extremity, due to the vaso-motor paralysis, tend to accelerate the pulse-wave velocity. The view is entirely unsupported, and the specious *prima facie* is disposed of by the consideration that the pulse-wave is distinct from the blood-stream, and that the velocity of the pulse-wave is so much greater than the velocity of the blood-current, that even if variations of the latter exerted their quantum of influence on the former, the effect would be inappreciable.

The opposite view—namely, that the pulse-wave is retarded under the conditions of vaso-motor paralysis, is supported by all the facts brought to light which have any bearing on the question. First, there is the accepted law in physics, that wave-impulses, communicated to a liquid flowing through a tube, are transmitted with a velocity proportioned directly to the rigidity of the tube. Under paralysis of the muscular coat, the artery becomes lax and yielding, and in obedience to the law just stated, would transmit the pulse-wave with diminished speed. Second, the arteries become more rigid with advances of age, and ample observation proves the rule that the velocity of the pulse-wave increases with age. Third, in low-toned arteries, evinced by low-pressure, high-sweep, and dicrotic pulse-traces, it has been observed that the pulse-wave travels slower than in high-toned arteries, evinced by high-pressure, low-sweep, and non-dicrotic pulse-traces; and, especially, in the marked arterial relaxation of adynamic conditions, the pulse-wave velocity has been found reduced. Moreover, as bearing directly upon the question at issue, the following case and experimental results are presented.

Ernest B——, aged thirty-four years, fell from a building, striking heavily the back of his neck and head against the ground. He retained his consciousness and power of speech, but suffered paralysis of voluntary motion and sensation of all parts below the neck. The next day imperfect motion, without co-ordination, and imperfect sensation returned to the upper extremities; this apparent improvement, however, was soon lost, and the case again presented the spectacle of complete paralysis of the entire body below the neck. The respiration was labored; heart's action apparently normal; pulse 60 to 80, full and regular; temperature ranging from 102° to 105° F. When the temperature marked 105° the pulse was full and not above 80. The man died on the fifth day after the receipt of the injury. The day before his death, by the kindness of Dr. A. B. Isham, whose patient he was, and from whom were received the above notes, the opportunity was given to make sphygmographic observations in the case. The respiration was then very labored, temperature 102½°, pulse averaging 62, full and regular, as shown in the tracings. (Fig. 125, Nos. 73, 74, and 75.) The tracings, especially the left carotid, are thrown out of horizontal line by the action of the respiration; they are also somewhat defective from extraneous causes; yet the basal points are correctly traced and the successions are truly measured and expressed.

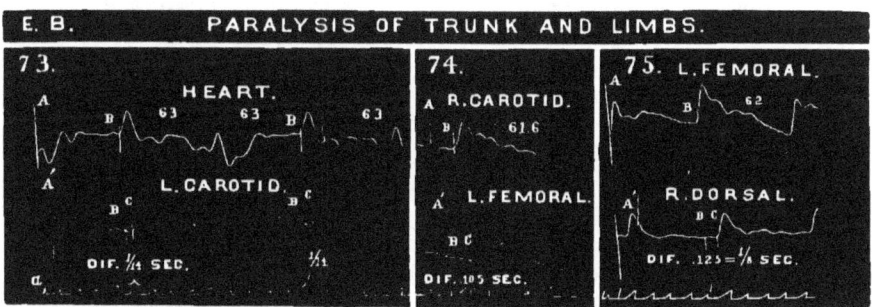

FIG. 125.

In this case the phenomena of vaso-motor paralysis were fairly declared, and the manifestations of pulse-wave velocity under the conditions eagerly looked for. The interval between the right carotid and left femoral is .105 second, that between the left femoral and right dorsalis pedis is

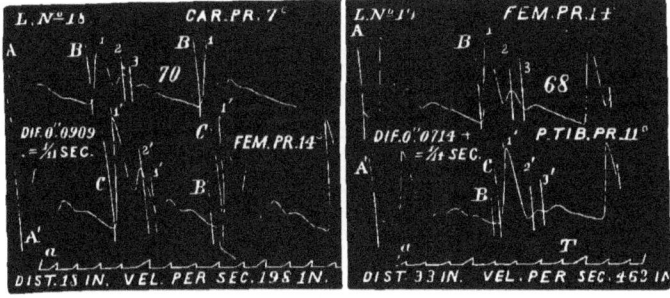

FIG. 126.

.125 second; the sum of which intervals gives .23 second as the time-difference of the pulse-wave between the carotid and dorsalis pedis. This is as slow a transmission of the pulse-wave as has been observed in unobstructed and non-aneurismal arteries of adults. The young man L., whose pulse-wave velocity was exceptionally slow, gave .0909 second for the carotid-femoral, and .0714 second for the femoral-posterior-tibial, (see Fig. 126, Nos. 18 and 19, here reproduced), the sum being .1623 for the carotid-posterior-tibial interval. So the retardation of the dorsalis-pedis pulse on the carotid in the paralytic case was about one third longer than that of the posterior-tibial pulse on the carotid in the case of L. These facts and considerations would seem to leave little room to doubt that the pulse-wave is retarded in its passage through an artery whose muscular coat is paralyzed.

From this study of the sphygmographic indications in aneurism, the following estimate of their diagnostic value seems fairly justified:

1. *Deformation* of the pulse beyond a suspected aneurism is of little, if any, value as indicating the presence of aneurism, but indicates rather the presence of arterial obstruction, which may originate from other conditions as well as from aneurism. Absence of such deformation is no evidence of absence of aneurism.

2. *Delay* of the pulse beyond a suspected aneurism is strong evidence of the presence of aneurism; and when arterial obstruction and vaso-motor paralysis can be eliminated, the sign is conclusive of aneurism. Absence of such delay is of no value as indicating the absence of aneurism.

3. In aneurism already diagnosed, deformation of the distal pulse indicates the presence of concomitant arterial obstruction.

4. In aneurism already diagnosed delay of the free distal pulse indicates one with free communications, large cavity, and yielding walls; while absence of pulse-delay indicates one with narrow orifice, or small cavity, or resisting walls.

CHAPTER X.

THE CARDIO-SPHYGMOGRAPHY OF TRICUSPID LESIONS—NEW INTERPRETATION OF FLINT'S MITRAL DIRECT OR PRESYSTOLIC MURMUR WITHOUT MITRAL LESIONS.

OBVIOUSLY the indications by the graphic method of lesions affecting the right auriculo-ventricular valve must be largely, if not wholly, negative, nevertheless their importance is sufficiently manifest. The value of negative evidence in diagnosis is well understood. Simultaneous cardio-arterial traces in left-side valvular disease give positive indications. Marked aortic stenosis, heavy aortic valves, mitral regurgitation, and aortic regurgitation, cannot exist without distinctive phenomena, which express themselves through cardio-sphygmography as diagnostic signs of the respective conditions. In right-side valvular disease such graphic signs are not obtained. Hence, when the question arises as to which side of the heart is the seat of the valvular trouble, the graphic records come in to decide.

Especially it is known how difficult it is by ordinary methods of examination to distinguish between mitral and tricuspid regurgitation. Indeed it is doubtful whether tricuspid regurgitant lesion has ever been positively recognized during life. Cardiac murmur, so much relied upon in the diagnosis of mitral lesion, is of little aid in the diagnosis of tricuspid lesion; nor are there any physical signs or general symptoms by which the latter can with any degree of certainty be differentiated from the former. The two following cases are in evidence and illustration of the important diagnostic aids cardio-sphygmography may afford toward the recognition of tricuspid lesion.

CASE I.—Mrs. Smith, aged fifty-nine years, came first under care in March, 1880. She was suffering from dyspnœa, cough, and general dropsy, which were found to be dependent upon organic cardiac disease. There were evidences of considerable enlargement of the heart, and a loud systolic murmur was audible over the cardiac region, plainly emphasized at the left apex, well heard to the left and in the back. The second sound of the heart was clear, and accentuated at the pulmonic site. Heart's action at times irregular; pulse somewhat accelerated, sufficiently ample; no albumen in urine. She continued under observation without material change of the physical signs, and died from failure of the circulation, May 8, 1881.

Simultaneous cardio-sphygmograms were obtained May 24, 1880, and February 12, 1881, of which Fig. 127, Nos. 91 and 92, are examples.

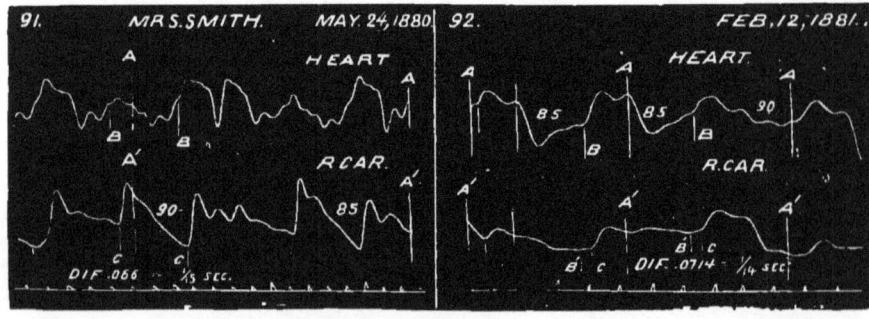

FIG. 127.

The post-mortem, concerning only the thorax, was conducted by Dr. A. B. Isham, in the presence, and with the assistance, of Dr. H. T. Lowry and the author. The pericardium contained a moderate amount of fluid. The heart was generally enlarged. Right cavities and great veins distended with blood, and the former contained firm fibrinous clots. Walls of right ventricle and right auricle greatly hypertrophied—dilated and thickened,—the former measuring one half inch in thickness. Venæ cavæ dilated; pulmonary valve normal; tricuspid orifice much dilated and valve incompetent, one of the segments bound down to the sides of the ventricle; the condition of the valve was such as must have permitted free regurgitation during life. Walls of left ventricle thickened to about three quarters of an inch, cavity not dilated; left auricular walls of normal thickness, and cavity of normal dimensions; mitral valve slightly thickened, but pliable and competent; aortic valves competent; texture of heart fatty and friable.

The ordinary signs and symptoms in this case pointed very definitely to mitral regurgitation as the predominant trouble. Especially the auscultatory phenomena declared this diagnosis, nor was any thing developed by the usual methods of examination to cast a doubt upon the correctness of this view. It was a case in which the physician, however skilful in cardiac diagnosis, confined to ordinary methods, would inevitably commit the error of assigning the regurgitation to the mitral orifice.

We will now examine the graphic records and see what light they might have shed upon the case. Allowing for the effects of the respiration and the irregularity shown in the middle of No. 91, Fig. 127, the only positive peculiarity of the cardiac trace is the high auricular wave. The carotid trace is free and ample in No. 91; shows feebleness in No. 92. The cardio-carotid interval measures $\frac{1}{15}$ second in No. 91, and $\frac{1}{14}$ second in No. 92, with pulse-rate in both 85 to 90 per minute. These intervals are strictly normal.

If free mitral regurgitation had existed, the cardio-carotid interval would have been abnormally long; it would have measured somewhere between an eighth and a quarter of a second. This important fact, independently established, still awaits the case to prove that the lesion may exist without the sign. In the present case the author thought he had found the exception to the rule, and stated to Dr. Isham that the sign had failed him, or the case was one of tricuspid instead of mitral regurgitation. He did not feel justified in arraying this simple negative sign against what appeared positive indications of mitral incompetency; and so was almost ready to concede in this instance default of the graphic indication, when the autopsy verified the integrity of the mitral and insufficiency of the tricuspid valve, attesting again the significance and trustworthiness of the graphic testimony.

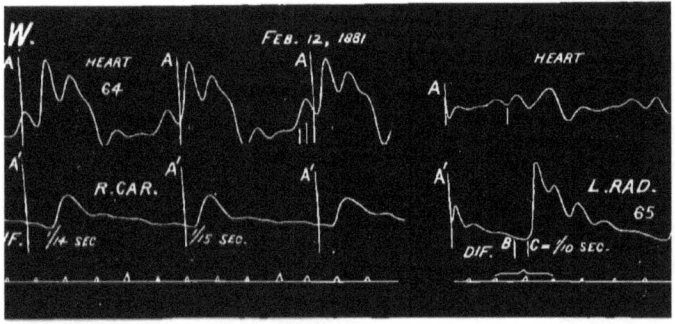

FIG. 128.

CASE II. John W., aged forty-eight years, a patient in the Cincinnati Hospital, suffering from heart disease, and whom the author was given the opportunity to observe by the kindness of the attending physicians. He was troubled with cough and dyspnœa, and had a double murmur at the base—the diastolic predominating—and a systolic murmur at the apex. Apex displaced downward and outward. Simultaneous traces, Fig. 128,* of the heart and carotid and

* Of the cardiac and carotid traces the former is excellent, while the latter is somewhat undeveloped in consequence of thickness of the tissues overlying the artery. Of the cardiac and radial traces the latter is excellent, while the former is undeveloped by a faulty application of the base. It is to be regretted that

of the heart and radial were taken February 12, 1881. The man died, and a post-mortem was made April 19, 1881. From the hospital notes and personal inspection of the specimen was gathered the following history: Marked dilatation of both ventricles and hypertrophy of ventricular walls. Right auricle dilated, and walls thickened. Tricuspid orifice contracted so as to admit only the tip of the little finger; valve thickened, and manifestly insufficient, as well as directly obstructive, during life. Pulmonary valves normal. Mitral valve normal. Aortic valves slightly thickened, and one contained a small calcareous plate; yet they preserved their normal form, and were evidently competent during life. Aorta dilated from near its origin to its bifurcation into the iliacs; especially the transverse and descending arch was dilated into a cylindrical pouch of considerable size; aortic walls thickened, rigid, and pervaded with calcareous plates varying from a quarter of an inch to an inch in diameter.

In this case no definite diagnosis had been arrived at; especially the tricuspid lesion had not suggested itself. The double murmur at the base was thought to indicate aortic obstructive and regurgitant lesion; and the systolic apex murmur, mitral regurgitation. A correct appreciation of the graphic showings would have demonstrated the absence of grave left-side valvular lesions. If aortic stenosis had been present, the pulse-trace would have been characteristically deformed, and not free and shapely, as shown, notably in the radial. If there had been obstruction at the aortic valves, or regurgitation at the mitral valves, one or both, the succession of the arterial pulses on the heart would have signalled abnormal delay instead of the intervals shown. If free aortic insufficiency had existed, the carotid and radial pulses, besides presenting the characteristic form of this lesion, would have appeared much sooner after the systole of the ventricle than

the last imperfection had not been corrected by a repetition of the experiment. However, the traces all show distinctly, and in proper position, the basal points, and so permit of correct determinations of the time-differences.

the time indicated. And further, if pure mitral* contraction had existed to produce such distress, ventricular systole would not have shown such development, nor the arterial pulse such fulness and amplitude.

Then, as the evidences of grave valvular disease were unmistakable, and the graphic records proved the integrity of the aortic and mitral, the localization was driven to the right-side valves, and attention to the signs of the graphic method would have secured a diagnosis which could not have been arrived at by other means.

The traces sufficiently indicate hypertrophy of the left ventricle, but they demonstrate also the presence of auricular hypertrophy. The latter information is furnished only by the graphic method; when either auricle is hypertrophied, there will be in the cardiac trace a correspondingly high auricular wave.

Finally, we find in the traces evidences bearing upon the arterial rigidity. The cardio-carotid interval being $\frac{1}{15}$ second, and the cardio-radial $\frac{1}{10}$ second, would give only $\frac{1}{30}$ second for the carotid-radial interval. This is less than half the average normal time-difference between these points in healthy adults, and as arterial rigidity is known to be the greatest of all factors increasing pulse-wave velocity, such unusual speed as is here signalled indicated that the arteries of the upper extremity partook of the rigidity so strikingly displayed in the aorta; and this rigidity extending in the innominate and carotid was the undoubted cause of the unduly short cardio-carotid interval. It is probable the pulse-wave velocity will come to be accepted as a test and gauge of the quality of the arterial walls as to their degree of rigidity or pliancy.

The diagnostic signs furnished by cardio-sphygmography go hand-in-hand with the evidences of auscultation and percussion, correcting, extending, and completing the latter, and give to cardiac diagnosis unparalleled precision and perfection.

---

* Opportunity for tracing mitral contraction has not offered, but experiments on the schema indicate that the pulse is abnormally delayed, also, in this condition.

New Interpretation of Flint's Mitral Direct or Presystolic Murmur without Mitral Lesions.*

Flint describes an interesting auscultatory phenomenon, namely, the occurrence of a cardiac murmur preceding and running into the first sound without mitral lesions, but always in association with considerable aortic insufficiency. He speaks of the murmur as *blubbering* in quality, best heard within the apex, and characterizes it as a *presystolic or mitral direct murmur*.

Although this murmur, it appears, has been recognized and pointed out alone by Flint, the correctness of the observation is not to be questioned. Discussion, however, is allowable regarding its interpretation and mechanism of production.

As just stated, our distinguished master refers the murmur to the mitral orifice, and terms it presystolic. The following is his explanation : " The left ventricle is filled with blood from the current passing from the auricle to the ventricle through an unobstructed orifice by gravitation, and, in addition, the regurgitant current from the aorta. As a consequence the mitral curtains are floated away from the ventricular walls and are not only approximated but in actual contact. These conditions existing, the auricle contracts and forces an additional quantity of blood into the ventricle. The mitral direct current passes between the valvular curtains, which are in apposition, and throws them into vibration." Is the explanation sufficient? It would seem the segments would not arrange themselves in apposition, because to do so would require a superior pressure on the ventricular side and the creation of a retrograde current towards the auricle. The fact is that during ventricular diastole the current is towards the ventricle, or possibly at a stand-still in the last moment before the occurrence of auricular systole. It is not conceivable that a retrograde current in the meantime could be established to close the segments, and from

* *American Journal of the Medical Sciences*, July, 1862. The same, April, 1882. The *London Lancet*, American edition, April, 1883.

the first moment of auricular contraction the direct current would keep them well open. The experiment of injecting water through the mitral orifice into the ventricle, and proving that the valve becomes closed, succeeds only because a brisk retrograde current carries the curtains into apposition. The conditions are not analogous to those of the living mechanism.

But even if we could conceive of the valve as closed in the manner assumed, the conditions would not yet be present for the production of a presystolic murmur. The curtains would be limp and passive and swayed noiselessly apart as the blood under the auricular systole passed between them into the ventricle. They would lack the requisite tension or stamina to permit of their being thrown into sonorous vibrations by the action of the passing current.

For these reasons the theory offered by Flint does not appear to us adequate to explain the phenomenon of a cardiac murmur before the first sound in aortic insufficiency without lesion of the mitral valve.

We will now proceed to give what we conceive to be the true explanation.

In normal cardiac action the first sound of the heart is heard a notable time after the beginning of ventricular systole. This is readily demonstrated by employing a cardiograph and stethoscope at the same time and giving attention to the play of the lever and sounds of the heart. It will be observed that the lever has nearly or quite completed its ascent when the first sound impresses the ear. The time of this ascent marks the interval between the beginning of ventricular systole and event of the first sound. The interval is measurable and may be stated roundly at about one fifteenth ($\frac{1}{15}$) of a second, which is long enough for a practised ear to appreciate.

Now, when the valves are perfect, the blood begins to escape from the ventricle through the aortic orifice about the time of culmination of the first sound, and a murmur generated at the aortic orifice would be heard with and following the first sound. This murmur would be counted, and truly, systolic. But when the aortic valves are permanently open, the blood would begin to flow through them immediately upon the beginning of ventricular contraction ; the consequence would be the generation of a murmur which would begin before and run into the first sound. This murmur, though also in reality strictly systolic, would strike the ear as occurring before the beginning of systole, because antedating the first sound of the heart.

Ventricular systole is an event which measures on an average .326 of a second, while the first sound of the heart, as appreciated by the ear, is of momentary duration. The first sound divides systole into a short anterior and longer posterior portion. The distinction here stated is not found in any of our classical works on the heart ; nor have any of our distinguished auscultators, so far as known, recognized a chronometric difference between the beginning of systole and the first sound of the heart. Without this recognition a systolic murmur beginning before the first sound would of necessity be referred to the mitral orifice and distinguished as a mitral direct murmur. With this recognition opportunity is given for the localization of such murmur at the aortic orifice, and distinguishing it as an aortic direct murmur.

That the blood begins to flow through the aortic orifice in aortic insufficiency immediately upon the beginning of ventricular contraction, instead of after a notable interval, as when the valves are intact, has been distinctly demonstrated by the simultaneous graphic method. It has been determined that while in normal conditions the carotid pulse follows the systole of the ventricle at an interval of one twelfth second, in aortic insufficiency the interval may not be more than one thirty-second second.

If scope permitted we could produce here simultaneous traces, from both man and the schema, in demonstration of this interval when the aortic valves are competent, and of its obliteration when they are open.

Under the cardiac conditions named all is favorable for

the development and definition of systolic aortic murmur in antecedence of the first sound. The enlarged ventricle and hypertrophied walls would suddenly start a volume of blood through the altered orifice, engendering vibrations, while the closure of the expanded mitral valve would give smartness to the first sound, and so materially aid the differentiation of the latter from the murmur already in progress.

Evidently the "blubbering" quality attributed to this murmur could be produced at the aortic as well as at the mitral orifice; and it is not unusual for an aortic murmur to be well heard at or within the apex. In the cases cited by Flint, in the first, "at the apex was a presystolic blubbering murmur. At the base of the heart was an aortic regurgitant murmur, which was diffused over nearly the whole præcordia." In the second case were found "an aortic direct and an aortic regurgitant murmur, both murmurs being well marked. There was also a distinct presystolic murmur within the apex, having the blubbering character." In a third case, recently reported, were four murmurs referable to the left side of the heart without mitral lesions. "The patient in this case had undoubtedly both a mitral systolic and a mitral presystolic murmur." In such a medley of sounds we appeal if it would be possible even for our author to decide that the murmur in question proceeded from the mitral orifice, independent of the preconception, that a murmur prior to the first sound is presystolic and necessarily mitral direct.

Other authorities, while accepting the soft presystolic murmur as signalling when present mitral obstructive lesion, do not admit the existence of a rough presystolic murmur under any conditions. The murmur under consideration they time as systolic. Flint likewise formerly considered it systolic. This goes to show how close must be this murmur to the first cardiac sound,—so close, indeed, that the superior and long-cultivated ear of Flint was required to first signal the distinction. A true presystolic or auricular systolic murmur precedes the first sound sufficiently long to be recognized as such without difficulty. It is true that it is more liable to be counted diastolic than systolic.

And yet, now that the distinction has been pointed out, we conceive it would not be difficult for the trained and attentive ear to appreciate this initial systolic murmur as precedent to the first sound. The contrast in the note of the murmur and note of the sound would help the discrimination. And we know that reduplication of the cardiac sounds is well perceived, although here two similar notes must be as near, at least, as the dissimilar notes of the murmur and sound under consideration.

Then accepting the fact as set forth, that a cardiac murmur preceding the first sound may be audible in certain cases with healthy mitral but insufficient aortic valves, our explanation of the phenomenon is, that the murmur is systolic and produced at the aortic orifice, and begins before the culmination of the first sound, because the blood is already flowing through the altered valves, throwing them into vibration, when the heart-note is heard.

The discovery and appreciation of the phenomenon by our illustrious author was a very natural result in view of his large clinical experience, and exquisite auscultatory powers, and the extreme care with which he investigates his cases. His interpretation was framed on the basis of the idea that the murmur was essentially mitral direct, and without reference to the positive results bearing upon the mechanism furnished by the graphic method. These results are too definite and significant to be neglected or ignored; and if we have succeeded in arriving at the true interpretation, it is because we have studied the question in the light of the results of cardio-sphygmography.

As a term is needed to designate this murmur, let it be named the *Murmur of Flint*.

LATEST MEDICAL PUBLICATIONS.

ALTHAUS. On Sclerosis of the Spinal Cord. Including Locomotor Ataxia, Spastic Spinal Paralysis, and Other System Diseases of the Spinal Cord ; their Pathology, Symptoms, Diagnosis, and Treatment. By JULIUS ALTHAUS, M.D. With nine illustrations. 8vo, cloth $2 75
"We commend it as an admirable compendium of knowledge on its subject."— *Medical News*, Louisville.

ANSTIE. Neuralgia and the Diseases which Resemble it. By FRANCIS E. ANSTIE, M.D. 16mo. 1 25
"The work treats instructively and entertainingly of an important subject. * * * Such an exhaustive treatise should not fail of a welcome."—*N. E. Medical Gazette*, Oct., 1885.

BULKLEY. Acné and its Treatment. A Practical Treatise Based on the Study of 1,500 Cases of Diseases of the Sebaceous Glands. By L. D. BULKLEY, M.D. 8vo, illustrated 2 00
"No one can pretend to treat acné with success unless he makes himself thoroughly conversant with Dr. Bulkley's work on the subject."—*New Orleans Medical and Surgical Journal*, Nov., 1885.

CORNIL and RANVIER. Manual of Pathological Histology. Part II., Vol. II. (Lesions of the Organs concluded). Second edition, illustrated. Translated, with approval of the authors, by A. M. HART. 8vo, cloth 6 00
Vol. II., Part I. (Lesions of the Organs), illustrated. 8vo, cloth 4 25
Vol. I. (Histology of the Tissues). 8vo, cloth 8 40

CORNING. Brain-Rest. By J. LEONARD CORNING, M.D. Second edition, revised and enlarged. 16mo, cloth 1 00
"Sensible and scientific, and cannot be too highly recommended to the attention of those troubled with insomnia and brain exhaustion."—*Pioneer Press*.

CUTLER. Manual of Differential Medical Diagnosis. By CONDICT W. CUTLER, M.D. 16mo, cloth 1 25
"This manual has decided merit, and will commend itself to every one engaged in the study of medicine. * * * The author displays rare skill and judgment in contrasting disease. His differentiation is clear, but not too sharply drawn, and displays extensive labor and research as well as practical knowledge."—*N. Y. Medical Journal*, Oct., 30, 1886.

DOWSE. The Brain and the Nerves: Their Ailments and Exhaustion. By THOMAS STRETCH DOWSE, M.D. 8vo, cloth 1 50
"Is readable, instructive, and suggestive."—*Medical Review*.

FARQUHARSON. School Hygiene and Diseases Incidental to School Life. By ROBERT FARQUHARSON, M.P., M.D., Edin. Cloth, 12mo 2 75
"The chapter on school buildings, school diet, and school diseases will well repay careful study, as they contain the fruits of much ripe experience."—*London Medical Record*, Sept., 1885.

FERRIER. Functions of the Brain. By DAVID FERRIER, M.D., F.R.S. Second edition, re-written, with many new illustrations. 8vo, cloth 4 00
This, though termed a second edition, is essentially a new book, having being almost entirely re-written, and embracing the results of new investigations by the author, as well as a critical survey of the more important physiological and pathological researches on the functions of the brain that have been published within the last ten years. The number of illustrations has been doubled, and the chapters devoted to the structure of the nerve-centres and the functions of the spinal cord have been much enlarged, so that the work forms a complete treatise on the central nervous system.

Medical Publications of G. P. Putnam's Sons.

FOX. The Influence of the Sympathetic on Disease. Illustrated. By EDWARD LONG FOX, M.D., F.R.C.P.; Physician to the Bristol Royal Infirmary. 8vo, cloth $6 00

FRIEDLAENDER. Microscopical Technology. A Manual for Use in the Investigation of Medicine and Pathological Anatomy. By CARL FRIEDLAENDER, M.D., Berlin. Trans. (by permission of the author) by STEPHEN YATES HOWELL, M.A., M.D. 16mo, cloth 1 00
"Concise, comprehensive, and scientific."—*Medical Record*, Dec. 26, 1885.

GARRATT. Myths in Medicine, and Old-time Doctors. By ALFRED C. GARRATT, M.D. 8vo, cloth 1 50
"The book is worthy of careful reading by both profession and laity."—*Lancet*, Detroit.

GRANGER. How to Care for the Insane. A Manual for Attendants in Insane Asylums. By WILLIAM D. GRANGER, M.D., State Asylum for the Insane, Buffalo, N. Y. 16mo, cloth 60

IRELAND. The Blot upon the Brain. Studies in History and Psychology. By WILLIAM W. IRELAND, M.D., Edin. 8vo, cloth 3 00
"Dr. Ireland's book will always be a standard authority. We hope it will have the success it unquestionably deserves."—*Edinburgh Medical Journal*, Nov., 1885.

KEYT. Sphygmography and Cardiography. Physiological and Clinical. By ALONZO T. KEYT, M.D. Edited by ASA B. ISHAM, M.D., and M. H. KEYT, M.D. With 128 illustrations. (*In preparation.*)

KITCHEN. Consumption : Its Nature, Causes, Prevention, and Cure. By J. M. W. KITCHEN, M.D. 12mo, cloth 1 25
"It will repay all who may spend the short time which is necessary to peruse it."—*Pacific Medical and Surgical Journal*.

—— Catarrh, Sore Throat, and Hoarseness. By J. M. W. KITCHEN, M.D. 16mo, cloth 1 00

KNAPP. Cocaine : Its Use in Ophthalmic and General Surgery. By Dr. H. KNAPP, and others. 8vo 75
"Full of interesting cases. We cordially recommend it to the attention of our readers. * * * It reflects great credit on the authors."—*London Lancet*.

LEFFERTS. Pharmacopœia for the Treatment of Diseases of the Larynx, Pharynx, and Nasal Passages. With Remarks on the Selection of Remedies, Choice of Instruments, and Methods of Making Local Applications. By GEORGE M. LEFFERTS, M.D. 16mo, cloth 1 00
"What is recommended in this work can be accepted as having been thoroughly tested."—*Canadian Practitioner*.

MACCORMAC. Surgical Operations. Part I. : Ligature of Arteries. A Short Description of the Surgical Anatomy and Modes of Tying the Principal Vessels. Ninety-three illustrations. By Sir WILLIAM MACCORMAC, Surgeon and Lecturer on Surgery, St. Thomas' Hospital, England. Cloth 1 40

MANN. A Manual of Prescription Writing. By MATTHEW D. MANN, M.D., Late Examiner in Materia Medica and Therapeutics in the College of Physicians and Surgeons, New York. Fourth edition. Revised, enlarged, and corrected according to the U. S. Pharmacopœia of 1880. 16mo, cloth 1 00
"An excellent little work, of value to the pharmacist as well as to the physician."—*National Druggist*, Nov., 1885.

MATTISON. The Treatment of Opium Addiction. By J. B. MATTISON, M.D. 8vo, cloth 50
"It is a clear, concise treatment which will interest the profession."—*Inter-Ocean*, Chicago.

Medical Publications of G. P. Putnam's Sons.

MEYNERT. Psychiatry: A Clinical Treatise on Diseases of the Fore-Brain, Based upon a Study of its Structure, Functions, and Nutrition. By THEODOR MEYNERT, M.D., Professor of Nervous Diseases and Chief of the Psychiatrical Clinic in Vienna. Translated (under authority of the author) by B. SACHS, M.D. 8vo, cloth $2 75
" We most earnestly urge our readers to put this work in their libraries as one that will prove indispensable."—*Quarterly Journal of Inebriety*, Jan., 1886.

MORRIS. How We Treat Wounds To-Day. A Treatise on the Subject of Antiseptic Surgery which can be Understood by Beginners. By ROBT. T. MORRIS, M.D. 16mo, cloth 1 00

OTIS. Practical Clinical Lessons on Syphilis and the Genito-Urinary Diseases. By FESSENDEN N. OTIS, M.D., Clinical Professor of Genito-Urinary Diseases in College of Physicians and Surgeons, New York. 8vo, cloth. Formerly $4.50; reduced to 2 00
" The work is very thorough in every detail."—*Medical Record*.

PARKER. Cancer: Its Nature and Etiology. With Tables of 397 Illustrated Cases. By WILLARD PARKER, M.D. 8vo, cloth 1 50
" Will prove of value to all who are interested in cancer."—*National Druggist*, Nov., 1885.

ROBERTS. Lectures on Dietetics and Dyspepsia. By SIR WILLIAM ROBERTS, M.D., F.R.S. Second edition, cloth 1 00
" We have read these lectures with singular pleasure, and feel that we have largely gained from their perusal. Every practitioner should carefully read them."—*Edinburgh Medical Journal*, Nov., 1886.

SEMPLE. The Diseases of Children. A Hand-Book for Practitioners and Students. By ARMAND SEMPLE, M.D. Cloth 1 75
" The book is a very fair presentation of its subject * * * and trustworthy. It is clearly written and systematically arranged."—*Medical Record*, N. Y.

STICKLER. The Adirondacks as a Health Resort. By JOSEPH W. STICKLER, M.D. 16mo, cloth 1 00

STUDENTS' AIDS SERIES. Paper, 25 cents; cloth, 50 cents.
Aids to Medicine.—Part III. (Double Part). Diseases of the Brain and its Membranes, of the Nervous System, of the Spinal Cord, and of the Ear. By C. E. ARMAND SEMPLE.
Aids to Medicine.—Part IV. Treating of Fevers, Skin Diseases, Worms, etc. By C. E. ARMAND SEMPLE. *In preparation.*
Aids to Surgery. By GEORGE BROWN, M.R.C.S.
Aids to Gynæcology. By ALFRED S. GUBB, L.R.C.P., M.R.C.S.
Aids to Obstetrics. (Double Part.) By SAMUEL NALL, B.A., M.R.C.P., London.

UPSHUR. Disorders of Menstruation (Students' Manual of). A Practical Treatise. By JOHN N. UPSHUR, M.D., Professor Medical College of Virginia. 16mo, cloth 1 25
* Contains many valuable hints."—*Medical Record*.
" Will prove a most useful little volume."—*Nashville Jour. of Medicine and Surgery*.

G. P. PUTNAM'S SONS, *in addition to their own publications described in this catalogue, keep on hand a full stock of all the current Medical and Scientific works. Prompt attention given to all orders or enquiries by mail. Having, in connection with their Branch House in London, special facilities, they are prepared to execute, at the lowest rates, all orders for foreign books and journals.*